Writing and Speaking
for Technical Professionals:
Communication Skills for Success

MARTIN S. RODEN
California State University, Los Angeles
Emeritus Professor of Electrical
Engineering

TERESSA MURPHY
International Business Machines Corp.
Vice President (retired)

Los Angeles, CA

Roden, Martin S., Murphy, Teressa
Includes bibliographical references
ISBN 978-0-9793487-3-0

Production Supervisor: **Raymond B. Landis**
Cover Design: **David McNutt**
Drawings by: **Michael Cox**
Editorial Consultant: **Kathy Landis**
Distribution: **Legal Books Distributing: (800) 200-7110**

Printed in the United States of America.

10   9   8   7   6   5   4   3   2   1

978-0-9793487-3-0

Preface

You can be an excellent designer, inventor, engineer, or researcher but if you are the only one who knows it, you will not be successful.

If you can't effectively express your ideas and convey the technical content of your work, you may as well not have done that work. It is absolutely essential that you be able to clearly and effectively communicate technical information.

All students must take writing courses as part of the English component of a technical degree. These courses generally cover use of vocabulary and grammar. They do not, however, deal with the aspects of communication unique to technical professionals. We simply don't write the way others do. As one example, we try to avoid complex "flowery" words in favor of simplicity and clarity.

The purpose of this text is to make you into a better technical communicator. Talk to anyone in industry, whether a recent graduate or a seasoned engineer, and you are sure to hear them say that effective writing and speaking are keys to success in the profession of engineering. The best idea or design is useless unless you can communicate the details to others.

Chapter 1 of this book sets the stage by giving examples of the importance of writing and oral presentations in industry. Then in Chapters 2 and 3, we explore the various types of writing from proposals to reports and patents. We do this first from the student's perspective and then from that of a technical professional. Chapter 4 looks at the essential tasks you must perform to begin any communication project, including assessing the needs and desires of the audience, and gathering information from a variety of sources including the Internet. Next, we devote an entire chapter to giving oral presentations. These are extremely important in industry, with most companies holding periodic (e.g., weekly) design reviews where you will have to present your progress. Chapter 6 is entitled "Rules and Tools", and may well be considered as necessary medicine to support this entire process. Within this chapter, we review grammar and also the essential elements of word processors. The final two chapters address the specifics of formulating a written document such as drafting the document, editing, adding biographies and graphics, and avoiding plagiarism.

The text contains many examples since we don't want to overwhelm you with rules. Our experience has taught us that most technical students don't look forward to taking a writing class. We don't subscribe to this philosophy at all, and believe that becoming an effective communicator can be satisfying and even a bit of fun.

The internet contains many excellent sources of information related to technical communication. The list of such sources is continually changing. Therefore, rather than include these in an extensive list of references, we have chosen to make the list available on the publisher's website. Go to www.discovery-press.com and click on **BOOKS** and then click on this text. You will find a variety of resources and links to useful websites. You will also find postings of examples relating to many of the topics in this book.

If you take this seriously and carefully read and absorb this material, we assure you that you will become a more effective communicator. Communication skills usually do not come naturally. You have to work at it.

Teressa Murphy

Martin S. Roden

Table of Contents

Preface

1. Communication Skills: Why Bother?

2. Student Writing

3 Professional Writing

Chapter 1

Communication Skills: Why Bother?

1.0 The Purpose of This Book

We will devote a short chapter to convince you that you should care about this subject. Most of your textbooks don't have to do this. Indeed, when you pick up a calculus text, you probably don't need a lot of justification for taking it seriously. You either know how important mathematics is to engineering, you trust your professors, or you want to get a good grade in the class. Because we all enter college knowing how to speak and write, it may not be as obvious to you that improvement is needed.

This book has been developed to help technical people and people working in technology (often referred to as "technology professionals") become good communicators both as students and as professionals. For simplicity, we sometimes refer to these people as "engineers," although the concepts apply to all of the applied sciences, or "technology professionals," to include a variety of functions in a technical or scientific organization.

The book first demonstrates the importance of writing and shows how effective communication will help you succeed in your career. We discuss various techniques for good writing and give students the necessary tools to begin a writing project. We also include exercises and examples that will help you practice and improve the processes required for effective writing and speaking. Incorporated into the textbook are guidelines so you can use this both as a text and as a handbook once you are in the workforce.

It is not intended to be a comprehensive English writing text, but to cover the topics most important to working professionals.

The topics of the book are:

- Importance of Communications
- Student Writing
- Professional Writing
- Beginning a Project

- The Writing Process
- The Editing and Revision Process
- The Final Document
- Grammar
- Plagiarism
- Presentations
- Meetings and Voicemail
- Additional Resources

Communication skills improve through practice, so each chapter contains exercises and case studies. Because most technical professionals work as part of a team, many of these exercises require students to work in a group.

Collaboration

The Basics

Who are you?

Understanding your likes and dislikes can help you determine the best speaking and writing processes for you. Most engineers would rather build things than do anything else. When they have spare time, they would rather play a computer game than read a book. Some engineers have no trouble speaking and seek out opportunities to express themselves. Others avoid speaking and would rather let others take the floor.

What is good writing?

Recognizing good writing helps to improve your own skills. When writing is sloppy, difficult to understand, or illogical, readers become frustrated and don't even want to read it. After all, the purpose of communications is to convey information, so good writing must have the following characteristics:

- It must be clear and concise.
- It must incorporate good grammar, punctuation and spelling.
- It must be appropriate to the audience.

Barriers to Effective Communications

There are several common barriers to effective communication. Following are the five most imposing ones:

1. Attitude

This book will help students understand why good communication skills are so important for success in the classroom and in the workforce. If you approach a communications project with a negative attitude and you don't think it is important, you are certain to do a sloppy job.

2. Lack of confidence

Just as you need a positive attitude, you need self-confidence to produce an effective result. By understanding the writing process and rules, students will gain experience and confidence they need to complete projects successfully.

3. Perfectionism

Perfectionism is the human tendency to regard anything that is not perfect to be unacceptable. Engineering is an exacting profession, so perfectionism is common. But in communications, perfectionism is a huge roadblock. A presentation or document doesn't have to be perfect – there's plenty of time for editing! This book presents techniques to improve writing over time and encourage a process of continuous improvement.

4. Pride of authorship

Because technical professionals often work as part of a team or write for a manager, students need to be prepared to have their work critiqued, edited, and changed. This book shows how a student should approach this process, how it works, and how it can help improve communication skills.

5. Procrastination

Students – and others – often put off starting work on communications projects for a variety of reasons. This book will help students plan and begin work in a productive and structured way.

Our goal is to help you overcome these and other barriers by learning and practicing good communication techniques. We also discuss the processes of writing and speaking in a clear, uncomplicated manner from an engineering viewpoint.

1.1 The Importance of Communication Skills

This book covers topics in verbal communications, both spoken and written. It will help technical students and professionals improve their communications skills by understanding and practicing straightforward methods of writing and speaking. By becoming better communicators, students and professionals will gain confidence and reap a variety of benefits such as improved grades and performance reviews.

Strictly speaking, "verbal communications" include spoken (or oral) and written communications. The word "verbal" means "of, relating to, or consisting of words," but has evolved to mean spoken, rather than written, communications. Be careful to ensure that there is no misunderstanding when you write, especially in proposals or contracts.

The book discusses communications necessary for students such as term papers, lab reports and resumes. It also covers communications skills needed by successful technical professionals for situations such as presentations, meetings, and documentation. Most technical professionals spend significant portions of their workdays communicating with others, so employers are eager to hire staff with excellent speaking and writing skills.

All of our lives we communicate with others in various ways to express our needs, feelings, and other information. The easiest and most natural way to communicate with words is by speaking. In fact, early humans were able to speak long before they developed the skill of writing by pressing symbols into clay or chiseling them into stone. It is no wonder, then, that writing can sometimes seem so difficult!

Early Forms of Communication

Because speaking is more natural than writing, improving speaking skills will help with writing skills as well. Many of the same processes are involved in both types of communications. In fact, when you write a speech and deliver it, you will need to be both a good writer and an effective speaker.

Speaking

For technical students, speaking well is not always recognized as an essential skill. Yet the first impressions we make upon meeting others in the workplace are based on appearance and speaking skills, starting with the first interview. As professionals, good speakers are able to communicate clearly with others and quickly establish the rapport needed to work together as part of a high-achieving team.

Following are a few types of speaking that you will do as a technical professional:

- Job Interviews – essential skills to obtain that first job and move into better positions
- Meetings – skills needed to be recognized as important contributor to the team, collaborate with others to work on a project, negotiate for project assignments
- Presentations – skills to educate others on project work or findings, convince others to support a project, or sell a product.

Writing

Many technical students disdain and detest writing. After all, if they wanted to write, they would have majored in English! They view the hard sciences as the only fields of merit and believe that anything else is just fluff. Because technical students don't have much time for reading, they may not know how to spell and use proper grammar, and writing becomes even more of a chore. They view a writing assignment as a prison sentence – a long time of suffering before they will be released. They sometimes procrastinate until they are right up against the deadline, then write like mad, leaving no time for reviewing, editing, and improving their work.

Well, in corporate America, good communications are essential. Investments in projects and technology are studied and discussed before decisions can be made. The people who can sell their projects using presentations, decision papers, and proposals are the ones who win business. Their audience is not necessarily composed only of people with a background in technology, but could include executives who studied finance, business or – gasp – English.

In today's professional environment we don't always interact with others face-to-face. First impressions are often based on writing skills exhibited in letters or emails. The proliferation of email as a primary communications tool makes writing skills more essential than ever. People just don't have the time to read poor writing.

The ability to express oneself in words is essential to convey concepts and information to others. When a concept is described, it becomes a reality. If it is only in the writer's head, it is just his or her thought. When the concept is understood and accepted, the author has created something of value. In fact, the basic product of a technical professional is an idea. Ideas are documented in reports, white papers, proposals, etc. that are an important part of a corporation's intellectual resources. Ideas – not unexpressed thoughts – can also receive legal protection to become valuable intellectual property.

Writing, then, is not something to just "get through" while you are in school and do well enough to get a passing grade. You will continue to write as you go through your career and will even enjoy writing when you are easily able to express yourself. In learning to write well, you will:

- learn to research a topic, analyze it, and express your ideas clearly.
- be able to provide an excellent writing sample for graduate school or employers.
- perform well in a professional environment and continuously improve your good skills throughout your career.

The benefits of good communications skills are numerous and varied– from getting good grades on papers in school to getting promotions at work. In fact, the benefits are enormous, especially when skills are established early and improved through practice – an essential part of learning to be a good speaker and writer.

1.2 What Do Others Say?

"Technical competence is essential, but the engineer's success, including in the areas of quality and safety assurance, sales and marketing, project management, and public policy, is dependent on effective communications through well-developed and utilized speaking and writing skills," says Patrick Natale, P.E., executive director of the American Society of Civil Engineers (ASCE). In fact, ASCE's policy on academic prerequisites for civil engineering licensure and professional practice notes that "the diversity of society is challenging our traditional views and increasing our need for interpersonal and communication skills." According to ASCE, the engineer of the future must be able to communicate with technical and non-technical audiences, convincingly and with passion, through listening, speaking, writing, mathematics, and visuals.

After decades in academia and industry, the authors of this text appreciate the importance of good communication skills – especially for engineers. But we also know engineers sometimes need to be convinced that these skills are worth spending time to develop. Some students are able to develop good speaking and writing skills in school while some technical professionals may be in the workplace for a few years before realizing that their poor communications skills are holding them back from advancement.

Punching the ticket to success

Most colleges and universities regularly survey their engineering graduates and the companies that hire them. In responding to these surveys, many indicate that they wish the school had placed greater emphasis on writing and speaking skills. Industry representatives echo this opinion and sometimes say that they would rather hire someone with excellent communication skills and an average GPA than someone with a very high GPA and poor writing abilities. Many people and organizations share the opinion that good communications are important for a successful career, no matter what the profession. A blue-ribbon commission founded by the College Board surveyed 120 human resource directors for its report, "Writing: A Ticket to Work…Or a Ticket Out, A Survey of Business Leaders," which was published in September 2004. The National Commission on Writing believes we must improve the quality of writing in our schools for students to succeed, both in school and in the workplace.

Following are some important conclusions of the report:

- People who cannot write and communicate clearly will not be hired, and if already working, are unlikely to last long enough to be considered for promotion. Half of responding companies reported that they take writing into consideration when hiring professional employees and when making promotion decisions. "In most cases, writing ability could be your ticket in . . . or it could be your ticket out," said one respondent. Commented another: "You can't move up without writing skills."
- Two-thirds of salaried employees in large American companies have some writing responsibility. "All employees must have writing ability…. Manufacturing documentation, operating procedures, reporting problems, lab safety, waste-disposal operations – all have to be crystal clear," said one human resource director.
- Eighty percent or more of the companies in the services and the finance, insurance, and real estate (FIRE) sectors – the corporations with greatest employment growth potential – assess writing during hiring. "Applicants who provide poorly written letters wouldn't likely get an interview," commented one insurance executive.
- More than 40 percent of responding firms offer or require training for salaried employees with writing deficiencies. "We're likely to send out 200-300 people annually for skills upgrade courses like 'business writing' or 'technical writing,'" said one respondent.

The survey found that technological advancement has increased the need for good writing skills. "With the fast pace of today's electronic communications, one might think that the value of fundamental writing skills has diminished in the workplace," said Joseph M. Tucci, president and CEO of EMC Corporation and chairman of the Business Roundtable's Education and the Workforce Task Force. "Actually, the need to write clearly and quickly has never been more important than in today's highly competitive, technology-driven global economy."

Following are some quotes on writing by a variety of famous people:

Mark Twain (19ᵗʰ century American writer): The secret of getting ahead is getting started. The secret of getting started is breaking your complex overwhelming tasks into small manageable tasks, and then starting on the first one.

Mark Twain: I didn't have time to write a short letter, so I wrote a long one instead.

Mark Twain: The difference between the right word and the almost right word is the difference between lightening and the lightning bug.

F. Scott Fitzgerald (20ᵗʰ century American writer): You don't write because you want to say something, you write because you've got something to say.

Edmund Morrison (20ᵗʰ century American author): Like stones, words are laborious and unforgiving, and the fitting of them together, like the fitting of stones, demands great patience and strength of purpose and particular skill.

Charles Caleb Colton (19ᵗʰ century English writer): Our admiration of fine writing will always be in proportion to its real difficulty and its apparent ease.

Joseph Pulitzer (19ᵗʰ century American journalist and publisher): Put it before them briefly so they will read it, clearly so they will appreciate it, picturesquely so they will remember it, and above all, accurately so they will be guided by its light.

David Hare (20th century English dramatist): The act of writing is the act of discovering what you believe.

Lord Byron (19th century English poet): Verbosity leads to unclear, inarticulate things.

Samuel Johnson (18th century English author): What is written without effort is in general read without pleasure.

Ivan Levison (21st century copywriter): When your writing is filled with detail, it has a lot more impact.

Ben Franklin (18th century author, printer, politician, scientist, inventor, and much more): Either write something worth reading or do something worth writing.

Peter De Vries (20th century American editor and novelist): I love being a writer. What I can't stand is the paperwork.

William Faulkner (20th century American author): I never know what I think about something until I read what I've written on it.

1.3 Exercises

1. Write a short essay (about 200 words) describing your own writing experience. Include a discussion of when you had to write your first paper as a college student. Also discuss how often you have to write any documents. Have another student in the class (or colleague at your place of employment) read and critique your essay.

2. a. Locate the website of any engineering professional organization (for example, if you plan to be a civil engineer, you might visit the American Society of Civil Engineering; for electrical engineering, you might go to the Institute for Electrical and Electronic Engineers and for mechanical engineering, to American Society of Mechanical Engineers).

b. Find the names of three people who have won major awards from that organization.

c. Conduct a web search to find out approximately how many papers/books/articles these award winners have written in the last five years.

d. Attempt to locate copies of two of these publications, read them, and write a short critique of them.

3. Obtain information about your favorite faculty member. Develop of list of that faculty member's publications in the past ten years. Locate two of these publications, read them, and make written observations about the writing style.

4. Interview a faculty member at your school or a colleague at your place of employment. In the interview, ask about the importance of communication skills. Write a short essay detailing what you learn from the interview process.

5. Try to obtain information regarding recent alumni surveys at your university (or any major university). Determine if questions were asked about the importance of communication skills. Write a short paper summarizing the alumni's views about writing and speaking well.

6. Locate a practicing engineer or other professional working for a technology organization (perhaps a friend or relative). Interview that person regarding the importance of communication in his or her workplace and get examples illustrating the need for good writing or speaking skills. Write a short paper summarizing your findings.

7. Write a fictional paper about an entry-level (fresh graduate) engineer who was not able to communicate well in the English language. Discuss what happened to this person during the first few months of employment and also discuss how the communication problem is likely to affect his or her career.

8. Locate a poorly-written instruction manual (it could be one you got with a recent purchase, or you can find such manuals on the web). These are easily found with inexpensive electronic devices (perhaps a digital

camera). Critique the manual by identifying its deficiencies and discuss the problems that could be caused by its poor writing.

9. Interview friends and colleagues to find a "horror story" concerning calls to customer technical support desks that have been outsourced to other countries. Write a short paper discussing problems that can occur in this situation. What do you think causes these problems? If you were in charge of technical support operations for such a company, what would you do to eliminate these problems? What if cost considerations prevented you from moving the help desk?

10. List five reasons why you believe that being a good speaker could help you progress in your career and five reasons why good writing skills are important to you. If you don't believe that communicating effectively is important to your career, list the reasons why you believe that improving your speaking and writing skills is a waste of time. Discuss these reasons with someone you respect in the field; write a short paper comparing and contrasting your views.

Chapter 2
Student Writing

2.0 Introduction

As a student, you will be called upon to perform a wide variety of writing tasks. In fact, you will probably complain bitterly about the proliferation of writing assignments. Rest assured that the faculty members are not asking you to do this because they are sadists. In fact, grading written assignments is far more labor-intensive than grading traditional engineering or technical ones. It is an activity that most engineering faculty members will tell you they were not trained to do.

So why do faculty inundate students with writing assignments? The simple answer is that faculty members know that you need communication skills to succeed in the profession. Alumni and employers send a clear and consistent message that the schools need to focus more on communication skills. Leaders in industry tell us that beginning engineers are often poor writers and speakers and are not equipped to work in their organizations.

Engineering has become highly complex. The day of Archimedes, who about 2,400 years ago ran down the block naked yelling "Eureka!" are over. What does this have to do with writing? Well, Archimedes had discovered the Archimedes Principle (named *after* he discovered it, not before) governing objects floating and displacing water. He did this while taking a bath, and was so excited that he went running down the street announcing his discovery.

Today, discoveries are rarely made by one person;

they're usually made by teams of people (and usually wearing clothing). Each person or team of people iterates on previous discoveries by enhancing or adding to those results. If an engineer cannot communicate the importance and details of the technical work, it may as well not be done at all.

In this chapter, we focus on the six most common types of writing performed by students. These are:

•Laboratory Reports
•Term Papers
•Project Reports
•Emails
•Resumes
•Letters to Prospective Employers

This list of items could certainly be expanded, but we have chosen these six since they represent a broad range of writing and are distinct from one another.

2.1 Laboratory Reports

The writing style of a lab report is formal and detailed. You should not say things like, "I hooked up the equipment and got good results." Remember that "good results" is a judgment call and your report should just give the facts. Also remember that the purpose of engineering labs is *not* to prove a principle. Indeed, the theory you learn in the classroom never represents a perfect and complete model of the real-world phenomenon. So please don't write anything like, "This proves the second law of thermodynamics," in your lab report. You have not proven this law, but have simply compared lab performance to what the law predicts.

Professional engineers often use the passive voice when writing technical papers and reports. For example, they write, "The problem was solved," instead of, "I solved the problem." The first sentence is called passive voice since "problem," the subject of the sentence, is not acting; it is being acted upon. Engineers can use the passive form if the acting party is not known or if the acting party has already been identified as the writer or a team. Additionally, some writers choose the passive form to avoid the overuse of "I" or "me" in a technical paper to make it sound more objective or modest. Generally, however, the active form makes writing more clear and dynamic.

Some universities ask students to use the active form in their lab reports. In this case, you would write, "Our team performed the measurements," rather than, "The measurements were performed." This is not always the case, so make sure you determine the style desired by your particular school or department. Note that the active voice answers the "who?" question, so is sometimes recommended for accuracy.

Most universities have their own specified format (template) for lab reports. Although there are variations from school to school, we usually find that the report contains the following major parts:

LABORATORY REPORT COMPONENTS
- Title page
- Introduction
- List of materials
- Instructions/Procedure (either paraphrased or copied from a manual)
- Summary of theory related to the topic
- Measurements taken (usually in tabular or graphic form)
- Discussion of results
- Conclusion
- References

The *title page* usually contains the name of the university and department, the course name and number, the name (and perhaps number) of the experiment, the name(s) of the students conducting the experiment, and the date.

The *introduction* clearly states the purpose of the experiment.

The *list of materials* is very important. You should list not only components and pieces of test equipment, but also serial numbers and other relevant details. This latter requirement may seem strange to you. However, one of the main purposes of the report is to enable someone else to duplicate your results. Suppose one of your pieces of test equipment was defective or poorly calibrated. Your report could help zero in on that instrument so it could be repaired or replaced.

Don't assume that the reader is a member of your class and has a copy of the lab manual. You should repeat the *instructions/procedure* from the lab manual. Many universities permit (even encourage) you to make copies from the original manual and include these as part of your report.

The *summary of theory* is optional. Not every school wants you to summarize the theory as part of your lab report. Even if it is not required, consider writing a few paragraphs summarizing the applicable theory related to the experiment. This will allow you to observe whether the experimental results seem consistent with the theory, and it will help you write the discussion of results.

The actual *measurements* are the real guts of the report. Much of a lab report should be prepared using a word processor. The measurements, however, usually are hand written and presented unedited directly from the procedure followed in class. You could add some presentation formats, such as graphing the data, but the original data should be part of your report. In some industries, engineers keep a bound notebook where pages cannot be inserted, destroyed, or moved. Then at the end of each day or week, someone else signs and dates the book. This could be important if your company applies for a patent and tries to document the original date of the invention.

Discussion of results is where you have a chance to shine as an apprentice engineer. Please, please, please NEVER write, "The results were good." "Good results" might have meaning if you were a newspaper writer reporting on a sporting event, but not when performing an engineering experiment. "Good" is a meaningless, unquantifiable adjective in the context of precise technical applications. Take the time to analyze the results and compare them to the predictions based on theory. Try to explain any discrepancies. Don't be disappointed if there are discrepancies. In fact, you should worry if all of your results agree with what the theory predicts. Remember that the theory learned in the classroom represents an attempt to model real-life phenomena as closely as possible without using an unmanageable number of parameters.

> Discrepancies do not mean that you made mistakes in the experiment. They may simply point out any shortcomings of the models.

Throughout this book, we will emphasize the importance of *introductions* and *conclusions*. The introduction sets the stage and tells the reader whether the paper is worth reading (thank goodness your instructor is required to read your report whether or not you have an interesting introduction). The conclusion is the final summation of your report. It should leave the reader convinced that you have accomplished your stated purpose. It can be as simple as, "We have explored the second law of thermodynamics and verified that it is an acceptable model of behavior in the lab."

On the other hand, it can be more extensive including the major results and conclusions of the experiment. As an example, you might write, "We have explored the second law of thermodynamics as it relates to this experimental setup. The temperature results were within 5% of those predicted by the theory. Because this variation is expected given the tolerances of the various components, we conclude that the theory leads to an acceptable model of behavior in the lab."

2.2 Term Papers

Term papers are reports written at the end of a quarter or semester covering a particular topic. They are sometimes referred to as *research papers*. In some classes, you will be assigned a topic, while in others, you are free to choose your own topic. The subject of your paper should be related to the material covered in the class.

> The major steps involved in writing a term paper are:
> - Choose a topic
> - Research the topic and develop a list of resources
> - Study the resources and develop your own notes
> - Write a draft of the paper
> - Edit the draft

In *choosing a topic* for your paper, you should not be too general. You are not writing a textbook on the subject. Examine your notes and textbook to locate specific topics that excite you and seem to have room for expansion. Perhaps an application is briefly mentioned in the text. Or maybe several topics from your notes can be combined, compared, or contrasted. Generally you want to think of the topic as a question you need to answer. Here are some examples:

- What are the applications of GPS in contemporary communication systems?
- How has nuclear energy developed relative to commercial applications?
- Are there similarities between the propulsion systems of commercial aircraft and of missiles?
- Do fuel cells have a future in consumer automobiles?

After you choose a topic, the challenge begins. You must *research the topic* to locate sources of information. Please don't fall into the trap of thinking Google™ is the answer to all of your prayers. Yes, the Internet is a wonderful source of information. However, you have to be a bit of a detective to sort purely commercial advertisements from websites with tutorial information. You should look beyond web search engines. Perhaps there are books in the university library that address parts of your subject. There may be papers in the numerous publications of the professional societies (e.g., ASCE, ASME, IEEE, Association for Computing

Machinery). You may be able to find articles in the popular literature (e.g., *Popular Science*, *Popular Mechanics*, *Consumer Reports*), although these will not be heavily technical in nature. Sometimes these references can point you to additional sources of information

Learn how to use abstracts to locate articles. Abstracts are described in detail in Section 4.3 of this text. You should become familiar with the *Engineering Index*, *Applied Science and Technology,* and *the Readers' Guide*. Major technical organizations may have their own abstract services such as *IEEE Xplore*.

Don't overlook people as a resource, especially people who are already working in the field. Most technical professionals love to talk about their work (unless it is sensitive or classified information, or they don't want you to know how little work they actually do). Perhaps there is a recent graduate you can locate who is working in a field related to your chosen topic, or maybe a fellow student has a part-time job in the technical industry. Maybe you can even get some information from your professors, though the course instructor certainly will not do all the work for you. Section 4.3 of this text explores various ways of researching a topic and gathering information.

Suppose you find one or more wonderful resources that precisely address your topic. It might be tempting to simply "cut and paste" to create your paper. But this approach not only represents plagiarism (trust us, your instructor can tell when something comes out of the literature), but it also deprives you of an opportunity to develop your own communication skills. No employer is going to pay you to reproduce materials that exist in the open literature, so understand now that you will be writing original papers throughout your career. We suggest you obtain the resources, and if you are able to make copies, mark up those copies (highlighter, marginal comments). Then write your own notes using these resources as a guide. Use your own words.

Let's now briefly outline the major components of a term paper. There will be variations based on the specific topic and instructions given by your professor or department. We will confine out attention to the major sections. These are:

> Components of a Term Paper
> - Abstract
> - Introduction
> - Body
> - Conclusion
> - References (bibliography)

The *abstract* is a brief summary of the paper. It is often read by those trying to determine whether they need to read the entire paper. Abstracts are usually not part of term papers unless those papers are unusually long. We discuss abstracts in detail in Section 7.1 of this book.

The next three sections, *introduction, body,* and *conclusion* represent the three broad categories of any communication project, whether it is a paper, an oral presentation, or even a letter.

The introduction attempts to draw the reader into the topic. To get the reader's attention, you need to understand your intended audience. Are you writing the paper for other students in your class? Or is the paper intended for people who are not majoring in engineering or a related field? You need to know whether you should assume that the reader of your paper has any prior knowledge of the subject. Your professor should indicate the audience to whom you are to write, but if the professor doesn't specify, you should ask about this. You generally want to gear your writing toward the less experienced members of your audience. The introduction should be as interesting as possible, and should give a hint about the scope of the paper. The introduction sometimes includes a brief history of the topic.

The *body* of the paper is the "guts" of your work. It contains the technical detail of the topic. It could represent as much as 80% of the entire paper. Here is where you introduce the technical content, including any graphs or other pictorial representations. As with the introduction, the format of the body of the paper depends highly on the intended audience, so you must determine that before you start writing.

The *conclusion* summarizes, wraps things up, and shows the reader that you accomplished your purpose. It sometimes tells about what comes next. That is, what future work needs to be done relative to the topic of the paper.

The conclusion should restate the theme of your paper and leave the reader feeling that you accomplished your objectives.

The three sections of the paper can simply be described as:

"Tell the reader what you are going to say, then say it, then tell the reader what you just said."

The *references* section is important for both educational and legal reasons. It directs the reader to more detailed sources of information. Readers of your paper may want to research the topic further, and your reference list can save them a great deal of time. From a legal (and ethical) point of view, you need to give credit to the various sources of your information. You could be accused of plagiarism (see Section 7.6) if you don't thoroughly credit sources of your information. References and footnotes related to direct quotes in your report are discussed elsewhere in this book (see Section 8.2). For now, we are talking about the resources you used to gather information. Reference lists are sometimes referred to as a *bibliography*. There are differences between reference lists and bibliographies, but today most people use the terms interchangeably.

Our discussion is meant to be general, and you must determine any special rules and requirements of your school or department before writing a term paper. The person who will give you the grade on the project is ultimately responsible for establishing and communicating the "ground rules". You are responsible for understanding your assignment at the outset, so be sure to ask for clarification if you have any questions. Remember the rule: "Don't assume"!

2.3 Project Reports

As an engineering student, you will likely be required to perform some type of capstone or senior design project. A major requirement of these projects is the final report. It is typically quite formal, and may have to follow a format dictated by your department or university. A project report can be thought of as a hybrid between term and lab reports.

If you are studying in the United States, your engineering program is probably accredited by ABET, the Accrediting Board for Engineering and Technology. ABET establishes criteria that programs must meet to be accredited. Although not explicitly required by ABET, senior design projects and reports can become an integral part of demonstrating that accrediting criteria are being met. Because this is so important, many schools want the final project report to reflect a number of the ABET requirements (which you can check out the ABET website; www.abet.org).

Unless your school provides other guidance, you should make sure to address the following seven topics:

1. What does your design accomplish?
2. What constraints were considered in your design?
3. What trade-offs were required and how did you optimize your design relative to the requirements of the project?
4. What alternative designs were considered?
5. How did you analyze the various approaches and select the best

---

Let's now expand on each of these topics.

1. *What does your design accomplish?* That is, what were your goals in performing the project? You need to clearly state why your design is significant enough for the reader to care.

2. *What constraints were considered in your design?* Every real-life design has constraints. Some may be as simple as a time deadline for completing the design or a budget for parts. Other constraints might include required size, power source, appearance, and safety.

3. *What trade-offs were required and how did you optimize your design relative to the requirements of the project?* In any real-life project, as some performance parameters improve, others are degraded. For example, in a radar design, as the system becomes more sensitive so it can detect smaller targets, the number of false alarms increases.

4. *What alternative designs were considered?* ABET uses the term "open-ended design." This means that a problem usually has more than one solution (unlike most problems in the back of a textbook chapter, where your answer is either right or wrong). Don't be afraid to be innovative in this process. For example, if you were designing a clock control system to automatically reset all clocks on campus when daylight-savings time begins, you might consider a digital communication system broadcasting from a central control point. An alternative is to hire students to run around campus and reset every clock by hand. Under certain constraints, this seemingly old-fashioned approach could be the optimum solution to the problem. Indeed, look at the "all hand car wash", where expensive automated equipment was removed and replaced by a line of workers with towels.

5. *How did you analyze the various approaches and select the best design approach for your project?* Here is where you can display your analytical skills as an engineer. Hopefully, you didn't randomly choose from among the various approaches to the design.

6. *How do your design results compare to objective and constraints?* You need to convince the reader that you made intelligent decisions about your final design.

7. *What safety or security issues need to be considered?* Every design contains safety or security issues. It's obvious that if you design an automobile, safety must be a consideration. In the case of cell phones, you need to worry about radiation effects or the fact that using a cell phone distracts attention from other activities (such as driving safely). Historically, engineers did not have to worry about these issues. The engineer could say, "I did what my manager told me to do." Today, more commonly, the engineer is responsible for broader implications of design.

Report Format Guidelines

If your school, department, or employer has a suggested format for your report, you must follow it. We suggest you use the following as a guideline.

The sections of the report consist of:
- •Abstract
- •Title Page
- •Acknowledgments
- •Introduction
- •Body
- •Conclusion
- •References

Lengthy papers might include additional categories such as a list of figures or a list of tables. These extra sections are usually not included in papers written for college assignments.

Except for *Acknowledgments*, these topics were discussed in the preceding section on term papers; the purpose and format of these are the same whether you are writing a term paper or a project report. Acknowledgments are different from references. They are usually reserved for recognizing institutions or individuals who have provided significant help in the project or report. It is very rare to find an individual who writes a significant technical paper without getting some form of support from other people. Do not view acknowledging support as a sign of weakness; it is really a sign of professionalism.

What about personal acknowledgments such as, "I want to thank my mother for enduring the pains of child birth to put me on this planet?" This type of acknowledgment is probably fine if you are receiving a major award. But it is not really appropriate for a technical report unless your parent or relative helped with the technical material in the report (sorry, Mom!). This type of acknowledgment is reserved for the dedication page, which is included in some books.

Whether you are writing a lab report, a term paper, or a project report, you will not do your best if you just sit at the keyboard and type the report, print it, and submit it. As we discuss in detail in later chapters, writing is an iterative process. You need to plan, draft, revise, revise, and then revise again.

In planning your paper, we recommend that you first develop an outline. Whether you create an outline from scratch or start with a standard format, you can determine the content by writing down the major topics you plan to cover and organizing them into the overall structure. You can do this on the computer or with pen or pencil and paper.

Once you have the outline, think about the content of each section. What kinds of things should the section cover? About how long should it be? What important details and analysis should you include in the body? Once you have thought about these items, you are ready for a rough draft where you simply get your ideas on paper. You don't need to worry about grammar, structure (sentences and paragraphs), section titles, or graphics (graphs or charts) at this point.

Then you should go through at least two revisions. While some people find that they can do all of this sitting at the computer terminal, we recommend a hybrid process between computer monitor and printed copy. Print out a copy of your draft with several spaces between lines for comments. Then critically read the copy and mark it up with notations about adding material or changing wording. Use your marked copy to enter the changes into the computer.

As you conclude the revisions, ask a friend or colleague to read the paper and make comments. When we read our own papers, we tend to see what we expect or want to see. After all, you wrote the paper to begin with so you certainly know what is in there. We can read an unclear sentence over and over again without realizing its shortcomings. Having a fresh set of eyes look at the paper is always helpful and highly recommended. At the very least, leave your paper for a time before you review it again and you will read it differently.

When you finish this entire process, you will have a paper that you can be proud of for many years to come.

2.4 Email

Some say that electronic mail − or email − is the most wonderful invention of the last century. Others say it is contributing to the downfall of society. Why is there such a love-hate relationship? Let's start with a list of some of the benefits and drawbacks of email.

What is good?
- Email is asynchronous − the transmitter and receiver don't need to be available at the same time.
- Email is inexpensive.
- Email is convenient.
- Email is fast.
- Detailed information in email is more accurate than that given orally.
- Email provides a written record.

What is bad?
- Email is impersonal.
- Email can be overused, causing people to avoid talking to each other – some issues are better resolved with a conversation.
- Email is so convenient that users often send emails without giving sufficient thought to content, significance, or accuracy.
- Email statements can sometimes be misinterpreted because you don't have the benefit of eye contact, tone of voice, or body language.
- Email is not completely reliable.
- Email leads to lots of junk mail (spam) since it is so cheap and convenient.
- Email opens up possibilities for innovative fraud techniques such as "phishing."
- Email can easily spread viruses throughout address books.
- Email is not private – most experts advise that users should consider email as public.

Do the advantages outweigh the disadvantages? You need to decide that yourself. But it is a very rare technical professional who does not extensively use email. Indeed, surveys show that some engineers spend up to 50% of their working time using email.

Email is less formal than other forms of written communication. It is often conversational. While you should be as accurate as possible, it would not make sense to spend excessive amounts of time editing an email message. If the receiver has trouble interpreting something in your transmitted email, a request for clarification can easily be sent. This is usually not possible with written formal papers.

The basic format of an email is similar to that of a memo which is detailed in Section 3.4. The header information starts with a "*To*" which is where you enter the recipient's email address. If your email software permits, this could be a group name, in which case the email will go to everyone in that group. Alternatively, you can send to more than one person by separating the addresses with semi-colons. With some email software, you can just put the person's name, and the computer will substitute the email address from your address book.

You don't need to specify who the email is from since your email address will be automatically attached to the email. You can configure most email software to add a "signature" to every email you send. You can set up the signature with as much information as you wish. It could be as simple as your name, or it could also contain a closing, your title, address and phone number.

Another header item is "*cc*" or "*copy*". The abbreviation "cc" stands for "courtesy copy", Any email addresses you list in this section will receive a copy of the item. The only difference between "*To*" and "*cc*" is that those who are sent copies are usually not the prime recipients of the email message so they often don't need to take any action based on its contents.

Suppose you wanted someone to see your email, but you didn't want those to whom it is addressed to know you have shared the copy. Perhaps you are emailing your supervisor with a complaint about something in the department. You don't want your supervisor to know you are sharing the complaint with others in the department. In such cases you can use "*bcc*" which means "blind courtesy copy". The addressees in this category receive the same thing as those in the "cc" category, but their addresses do not appear at the top of the email.

Another use of "*bcc*" is to protect people's privacy and email addresses. Suppose you have solicited donations to a particular political cause, and a number of your fellow workers gave money to the cause. You want to inform them of progress on the issue, and decide to email updates to the donors. If you addressed the email to those who donated, everyone who received it would see the names of the other people, and this may compromise their privacy. If instead you blind copy each of the donors they will not get to see the other addresses (Note: Most email programs require that at least one address appear in the "To" field, so, if necessary, you can address the email to yourself). One more aspect of bcc is that those who are blind copied will not receive any replies. If someone who gets your email chooses to send a "*reply to all*", this will only include those in the "*To*" or "*cc*" category.

Finally, there is a field labeled "*Subject*". You should choose a very short descriptive title for your email. Be cautious of spam filters.

Titles such as "Important", "Critical information", or "Don't miss this" can easily trigger spam filters and your intended recipient may never see your email. Most people are very busy. Some get so much email, they scan the subjects and decide which to read and which to delete based only on these several words in the subject.

This takes us to the body of the email. You need to be as brief as possible, and have any critical information at the beginning. Let's illustrate this with two examples:

To: Members of design group 17
From: Your boss
cc: Your boss's boss
Subject: Important

Dear members of design group 17,
As you probably are aware, our company has been purchased by a large capital investment group based in Korea. They have replaced a number of key individuals in upper management. You have probably noticed that my office has moved and I now report to a new laboratory manager. The new management conducted a thorough review of our operations, and brought in expert consultants on organizational strategies. They worked for ten months and have issued an extensive report. The report contains 14 recommendations for changes in the way we operate.
Your job is affected by all of these changes. We need to become informed of the new directions the company is taking and the new goals and objectives. We will be meeting next Wednesday at 10AM to discuss this. This is very important and you will learn a lot about this reorganization.
Sincerely,
Your boss

To: Members of design group 17
From: Jane McDonald, Supervisor
cc: J. Takamatsu, Laboratory Manager
Subject: Meeting, Wednesday, 10AM

Mark your calendars to attend a critical meeting next Wednesday at 10AM. The purpose of the meeting is to share ideas regarding the new company reorganization and the effects this has on our department. Please inform me if you will be unable to attend.

What's wrong with the first email? The most critical problem is that the important meeting information is buried in the second paragraph and does not stand out. Even the subject of the email does not convey any important information and could even be captured by a spam filter. The memo contains too much information. The "Dear members" salutation is not necessary nor is the "sincerely" at the closing. Although you should have a signature line (or even detailed contact information) at the end, emails are normally don't require closing phrases such as "sincerely" or "yours truly".

In contrast, the second email is brief and to the point. It highlights the important information in the subject line and the body of the note by requesting that the recipient attend a meeting at a specified time. The reader is not required to work hard to find out what he or she is expected to do. If you keep your emails short and easy to read, the recipients will be grateful and will want to read them. If you make sure that important information is highlighted, the reader will know what is expected.

We conclude with some brief comments on email etiquette and ethics. If you are using a computer at your university or place or employment, you should respect the fact that someone else is paying for the service and hardware. It is unrealistic to think you would never use an office or school computer for personal business. If you have a critical need to send a brief personal email and it cannot wait until you are back at your home computer, most supervisors would not complain. But when you spend a lot of time doing personal things on someone else's computer, you are opening yourself up to serious consequences. The courts have decided that the owner or a system (your university or your employer) has the right to monitor computer use by students and employees. Sending personal emails (including forwarding jokes you receive) could result in reprimands or other unpleasant consequences. Serious infractions (e.g., running a business on an office computer or distributing pornography) could result in dismissal or legal action.

True Story #1: A student received a pornographic email depicting photos of underage children. The student was shocked and disgusted, and forwarded the email to a friend asking the friend what to do. The university detected that email and federal law officials showed up at the student's dorm accusing him of distributing child pornography.

True Story #2: A student sold a bicycle on the internet. In the process, the student exchanged emails with prospective buyers using the school's computer. The student was brought up on charges with the explanation that all students sign a statement saying they will not use school computers for commercial business.

With all the options for personal communication devices, there is really no excuse to abuse office or university computer use rules.

2.5 Text Messaging

With the proliferation of handheld communication devices (e.g., cell phone and PDAs), text messaging has become a popular augment to email. The term is applied to the process of sending text messages between mobile communication devices, although text messaging can also be used with desktop computers or telephones. The messages are typically short – up to about 160 characters including spaces.

Text messaging is slower than voice messaging because of the time needed to type the message, sometimes on a very small keyboard. It is not a substitute for email because messages are limited in size and normally don't contain attachments. Nonetheless, this form of communications, which may have started with teenagers sending romantic messages, has now become a staple in industry.

The key to effective text messaging is to be brief. Since the number of characters is limited and most display screens are small, you must convey your message in about 30 words or less. While we encouraged you to be brief in writing email messages, that encouragement becomes an absolute necessity with text messaging. That is why the abbreviations we discuss next are so important.

Text messaging uses its own set of acronyms to depict standard phrases. This is done to save typing time and to keep messages within the constrained length. We now summarize the most common acronyms but a more complete list can be found at the following website:

http://www.webopedia.com/quick_ref/textmessageabbreviations.asp :

BRB	Be right back
BTW	By the way
FYI	For your information
IMHO	In my humble opinion
...	Place asterisks around a word for emphasis
POS	Parent over shoulder (or another phrase with street language describing a defective or unsatisfactory item of equipment such as a lemon car)
:-)	Happy face
:-)))	Happy face with double chins
;-)	Winking happy face
8:-)	Happy face wearing glasses
ABC	All upper case indicates yelling

2.6 Resumes

A resume is a formal summary of important information about you and your qualifications for a professional position. You will need to submit a resume when you apply for employment. Resumes should always be accompanied by a cover letter, as discussed in Section 2.7.

The resume is not intended to be your life story. Most entry-level applicants (your first real engineering job) should limit their resume to one page. The people to whom you submit the resume are very busy and often have to review a large number of resumes. You must make it as easy as possible for them to do narrow the list of applicants to a small group of finalists.

While the information required in a resume is fairly standard, the actual layout can be flexible. You will need to make decisions about the use of titles, bold print, and italics as well as the wording of the resume. Plan to spend enough time developing and reworking your resume since a professional-looking resume will get a much more positive response than one that has errors or appears to be carelessly prepared.

One size does NOT fit all. You should research the company to which you are applying and customize your resume for that job function and your perception of the company needs. An experienced recruiter can easily spot the "shotgun" approach (when a student sends out hundreds of standardized letters and resumes hoping to get at least a few hits).

Don't use jargon and abbreviations unless they are universally accepted and understood. For example, in referring to any previous work experience you have, avoid terms and abbreviations that will not be familiar to the recruiter. Be as concise as possible.

A resume is not a letter, so it is acceptable to have fragmented sentence and phrases. Just be consistent and clear. Avoid getting personal or informal – that is, don't use phrases such as, "I worked at McDonald's for the summer."

Spell-check your work and have someone else check it for errors. A sloppy resume is almost certain to lead to a rejection for the job.

Sections of a Resume.

Although some variation is acceptable, most resumes contain the following major sections:
- Contact Information
- Career Objective
- Education
- Work Experience
- Honors and Activities
- References
- Other

Your resume should start with contact information, and objectives usually follow. But the placement of the other sections can vary according to your strengths. Normally a student resume lists education first since that is what recruiters are most interested in seeing – and it is usually the student's strongest area. Most college placement offices have a standard format for resumes and require the student to place his or her education first. After all, recruiters who are visiting the school are interested in students who have completed a particular course of study. However, if you are tailoring your resume for a particular position and have considerable relevant work experience, you may wish to place your experience section before your education. Of course, an experienced professional should list work experience before his or her educational background.

Contact information is placed at the very top of the resume. It can be left justified, right justified, or centered – try each format and choose the one that makes the entire page look the best. The listing must consist of your full name, your address (if at temporary school address, give both permanent and campus addresses), phone number(s), email address, and web address if you maintain a personal URL which will not embarrass you. In giving phone number(s) and email addresses, make sure it will be convenient for the recruiter to contact you. A busy recruiter will not spend a lot of time trying to track you down.

Career Objective is usually a short section that can have a variety of titles. You can call it, "Career Objective", "Career Goals", or just simply "Objective". You can use either phrases (e.g., A position related to structural analysis and design) or something closer to full sentences (e.g., Seeking a position related to structural analysis and design). Your statement should be broad enough to cover the kinds of positions the recruiter is trying to fill, but not be so broad as to give the impression you don't know what you want. Avoid generalizations such as, "Seeking a position to apply skills acquired in the BSME program."

Education represents your major accomplishment prior to entering the profession. It should receive prominent placement in your resume. Education sections, like experience sections, are usually placed in the middle of a resume, somewhere between the objective statement and the **Honors and Activities** section.

Should you put your GPA on the resume? If it is impressive (say over 3.0), by all means put it there. And what GPA should you use (overall, major, specialization)? Choose the one that makes you look best and that you could defend without breaking into a cold sweat if challenged. If you have gotten only one "A" in your college life, say in Strength of Materials, please don't say you have a 4.0 in your major and then respond that your major is Strength of Materials. This could certainly backfire with the recruiter forming the impression that you are dishonest or deceptive.

While **Work Experience** is not a requirement for entry-level applicants, if you have been fortunate enough to gain such experience, it is a definite positive. Even non-technical experience (e.g., a part-time job at McDonalds) shows a level of maturity and ability to manage your own finances. Of course if you have any technical experience (e.g., a summer internship), place that first. This goes against the general rule of listing experiences in chronological order. If you have significant technical experience, then you may wish to eliminate the "McDonalds" jobs, or perhaps follow the technical listings with a general comment such as "held various part-time jobs to support myself while taking classes." Don't forget any volunteer work you may have done – it's very important to the recruiter. Just remember to always place relevant experience first.

Example of Employment History

Summer 2008   Engineering Intern, Boeing Systems. Assisted in design
                 of new aircraft using various CAD/CAM software
                 packages

2002-present   United Parcel Service, Washington. Worked as Shift
                 Manager, overseeing sorting process and planning
                 deliveries.

Recruiters want some assurance that you can work as a member of a team. A straight-A student who studies alone, never takes part in any clubs or organizations, and cannot interact with other students is not what most recruiters are looking for. They would rather see the B student who has been an officer in a professional organization, taken part in extracurricular activities, and has demonstrated the ability to be a team player. You can highlight these activities in the **Honors and Activities** section. Start with honors, which could include the dean's list, scholarships, awards, or membership in honor societies. Activities include all clubs, and particular emphasis should be given to any experience you have as an officer in an organization.

References are very important when seeking employment. Most prospective employers will contact your references to gain insight into your strengths and weaknesses. Your references can include professors from whom you have taken courses and supervisors to whom you have reported. Of course you should pick people who you think will say positive things about you. It is always best to ask the person's permission first. In the case of faculty members, we suggest you give them a current resume and thank them for agreeing to act as a reference. The general practice is to indicate on the resume that references are available on request. You can then tailor your list of references to match them to the employer and job responsibilities

Other possible sections of a resume depend on your experience. Some students include a "**Skills**" section where they list computer and foreign language competencies. Use your judgment as to whether a skill will distinguish you. For example, if you list a section on computer skills and only include MS Word and Excel, it will not get any points for you (any high school student can do that today). If you are applying for a position that may require a security clearance, you should state your citizenship status.

There are many places for help in preparing your resume. If you are a student, you can visit your campus career center. In addition to providing counseling and publications, many career centers conduct formal classes and workshops in resume writing. You should start working on your resume before your final year of school to help you identify areas that you can enhance before you start applying for permanent employment. Having a resume can also come in handy if you want to apply for employment or a special project while you are still in school.

Although there are many acceptable ways to construct a resume, let's take a moment to look at a hypothetical example[1]. This is a relatively good resume for someone who is just graduating with a master's degree; let's call her Donna Jones.

DONNA JONES
710 East Jones Drive #8   626-555-5555
Pasadena, California 91101
       2dlj@4link.net

OBJECTIVE   Seeking position in the field of electrical
                       engineering with emphasis in systems engineering

HIGHLIGHTS   * Strong background in computer hardware,
                       computer Communications, and systems
                       engineering
                       * Extensive web design experience
                       * High level of work ethics, team player, detail-
                       oriented, effective problem solver, excellent
                       communication skills

EDUCATION   **M.S. in Electrical Engineering** 06/2007

---

[1] Adapted from materials supplied by Ms. Joanne Martel of the California State University, Los Angeles Career Center.

B.S. in Electrical Engineering 03/2005
California State University, Los Angeles

COMPUTER SKILLS	**Languages:** BASIC, C/C++, Assembly, Java

Languages: BASIC, C/C++, Assembly, Java
Systems: DOS, Windows 95/98/NT/2000/Me/XP, Macintosh OS 7.1+, Sun Workstations
Software: MS Office, WordPerfect, Microsoft Exchange, Lotus, Filemaker pro, Adobe Photoshop, MATLAB
Other: Proficiency with internet, browsers, computer networking, Telnet, FTP Verilog, SPICE, PSPICE.

EXPERIENCE

Web Developer   06/03-Present
Charo's Barbeque Restaurants, Covina

* Design, test, implement and maintain web site for restaurant chain with online mail order business.
* **Wrote and regularly update** user manual.

Intern   06/03 – 09/03
NASA Undergraduate Research Program, Pasadena

* Worked with the computer architecture team on a NGST test bed
* Analyzed the parallel processing and fault detection on DSP processors
* Surveyed channel models for designing digital image transmissions

Sales Consultant   03/02-05/03
Planetlink, Inc., South Pasadena
* Sold internet services and created new customer accounts
* Exceeded sales quota 70% of the time

HONORS & ACTIVITIES

Ada I Pressman Memorial Scholarship Recipient
President, Abacus Computer Society

Member, Institute of Electrical and Electronic
Engineers
Member, Society of Women Engineers
Member, Toastmasters International

REFERENCES   Available upon request.

This example resume contains the necessary sections beginning with contact information. This student has chosen to include a short section on "**Highlights**" which consists of three bulleted statements summarizing her skills and qualities. The "**Education**" section contains the essential information including the dates of the degrees. If she had an impressive grade point average, it would probably be a good idea to include it. Because she possesses a reasonable amount of experience, the "**Experience**" section is longer than one normally finds for a recent graduate. Notice that the "intern" section is describing an on-campus research project, which is perfectly appropriate.

We conclude this section by presenting two examples of resume templates. The first is a very simple entry-level template which can be used by a recent graduate.

Standard word processor programs such as Word contain templates for various types of resumes. The final example shows the template for a traditional chronological resume. You simply load the template and then type your own information for each entry. As you type, the Word prompt will be automatically replaced by what you type.

Example 1: Entry-level resume format

Your Name

Objective

Enter objective here

Education

Degree (both completed degrees and
expected degrees)
Dates, institution

Awards

List any awards of honorary
memberships

Work Experience

Dates, job titles, company, brief
description of responsibilities

Volunteer Work

List any volunteer work. If none,
eliminate this section.

References

Either list names, positions, contact
information, or say
"Available upon request."

Example 2: Chronological resume (traditional design) template from MS
Word:

NAME

[STREET ADDRESS], [CITY, ST ZIP CODE], [PHONE NUMBER], [E-MAIL ADDRESS]

OBJECTIVE

[Describe your career goal or ideal job.]

EXPERIENCE

| [START DATE] TO [END DATE] | [Company name] | [City, ST] |

[JOB TITLE]

* [Job responsibility/achievement]
* [Job responsibility/achievement]
* [Job responsibility/achievement]

| [START DATE] TO [END DATE] | [Company name] | [City, ST] |

[JOB TITLE]

* [Job responsibility/achievement]
* [Job responsibility/achievement]
* [Job responsibility/achievement]

| [START DATE] TO [END DATE] | [Company name] | [City, ST] |

[COMPANY NAME]

* [Job responsibility/achievement]
* [Job responsibility/achievement]
* [Job responsibility/achievement]

| [START DATE] TO [END DATE] | [Company name] | [City, ST] |

[JOB TITLE]

* [Job responsibility/achievement]
* [Job responsibility/achievement]
* [Job responsibility/achievement]

EDUCATION

| [Dates of   attendance] | [School name] | [City, ST] |

[DEGREE OBTAINED]

- [Special award/accomplishment or degree minor]

REFERENCES

References are available on request.

-
-
-

2.7 Letters to Prospective Employers

What Is a Cover Letter?

A cover letter accompanies your resume when you apply for a position in an organization. As we just discussed, your resume contains information about your background, qualifications, and skills. A cover letter allows you to point out your most significant and/or relevant accomplishments and skills. You should personalize each letter to the position and organization and introduce yourself. The cover letter also provides a sample of your writing skills. So you should send one with every resume you submit. The cover letter addresses three major questions:

- Why you arc contacting the company?

- What distinguishes you from the other applicants?

- What do you hope happens next?

Your letter should begin by explaining why you are writing. It could be as simple as, "I am responding to your advertisement in the career newsletter at XYZ University." Or it could be more specific. For example, "I know ABC Corporation is a leader in GPS navigation devices, and I have a strong interest in building a career around this exciting subject."

Next you need to draw the recruiter's attention to your most significant accomplishments. It could be something like, "Please note from the resume that I have achieved a 3.5 grade point average in my studies

condense the letter to one page by eliminating some wording or, in a pinch, widening the margins a bit. Type size should never go below 8 points, but 12 point type is customary. Word processors normally give four choices for justification (spacing within a line of type). Standard typing using *left justification* where the left ends of the lines fall directly under each other, but the right ends are "ragged". The opposite of this is *right justification* where the right ends line up but the left are jagged. You can choose *center justification* where each line is centered between the margins and both left and right ends are ragged. The fourth choice is *full justification* where the spacing between characters is adjusted to make each line fill the entire space between left and right margins. Choice of justification type is a matter of preference. Letters usually don't use justification for the right margin (lining up the end of each line). Doing so makes the letter look too formal.

Leave one or more spaces at the end of the text, and then insert your *salutation*. The standard form is either "Sincerely" or "Sincerely yours." As with the date, this could be placed either on the left or right of the letter. Some people still use "Yours truly," although this is declining in popularity as it is sometimes viewed as too personal.

Then skip about 4-5 spaces and type your full name. After printing the letter, you will sign in ink in the space above your typed name. Sign neatly. Once you are famous or become a doctor, you can develop a distinctive and unreadable signature.

Following the typed name, skip a line and then type "resume," which indicates that a resume is enclosed with the letter. If you clearly state in your letter that a resume is attached, this last line is optional. However, it could be helpful if someone else is opening the mail and might overlook it.

The following page shows an example of a cover letter.

Jane Smith
2345 4th street
Cincinnati, OH 30508
(616) 555-1212
jsmith@znet.com

June 25, 2012

Michael Roberts
Director of College Recruiting
ABC Manufacturing
1234 South First Street
Arcadia, CA 91007

Dear Mr. Roberts:

I am responding to your listing in the ABC University Career Center job guide. I am very interested in a position in your GPS research and design area. I am a very hard worker and-goal oriented. My resume is enclosed showing my education, experience, and skills.

My resume shows a strong work ethic and the ability to balance competing demands. I have maintained a 3.1 grade point average and have been on the dean's list four times, even though throughout my college years, I have had to support myself. I am also active in ASME and was elected to the position of recording secretary of our chapter last year.

My work experience has ranged from the very routine (McDonald's) to a summer internship at DEF engineering. In the internship, I assisted a group of engineers in model testing and calibration. I learned a great deal during that summer and developed my team skills.

I am available for an interview at your convenience. The best way to reach me is using the email at the top of this letter. Alternatively, the telephone has an answering machine, so if you do call I will return your call quickly. I look forward to hearing from you.

Sincerely,

Jane Smith

Resume

2.8 Exercises

1. Review three lab reports you have written. Discuss, in writing, at least four ways you could improve these reports.

2. Evaluate the most recent term paper you have prepared. Pretend that you are the instructor of the course. Grade your term paper and make suggestions for improvement.

3. Pick one of your classmates to be your team member. Then locate one recent paper each of you has written and exchange them. Critique the other person's paper and suggest improvements. What did you learn about your writing? What did you learn about editing another person's work?

4. Gather together the last five email messages you have received. Evaluate their effectiveness using the principles learned in this chapter. What suggestions can you make about improving them?

5. Look through each of the last month's email messages and pick the three most effective and the three least effective emails. Write a paragraph about each of these six emails describing why you think it is good or bad.

6. Pretend that you are a project leader of a team of ten students preparing for a design competition. You want to hold a meeting of your group and are trying to decide when to do so. Draft an email message to your group members asking them to help choose the day and time for the meeting. Then critique your own email message and suggest at least five ways to improve it.

7. Create (or update) your resume using the principles you learn in this chapter. Then pretend you are the interviewer viewing this resume for the first time. What kinds of questions would you want to ask?

8. You are applying for a summer job at XYZ manufacturing, a company that designs and manufactures automatic scoring devices for bowling alleys. Draft a cover letter to apply for a job in the design department.

Then have two of your classmates critique the letter for you. What do you think about their comments?

9. Do research on the web to select three companies that you think you would like to work for. Then prepare a cover letter and resume for each of these companies. Critique your own work and edit your letters.

10. What do you think about the importance of good writing as a student? Do you have many opportunities to practice your writing or improve your skills? Think about the time you spend writing and estimate the percentage of time that is spent on each category: reports, term papers, emails, marketing letters, etc.

Costs: Detailed costing for all equipment and labor. Costs can be given for firm fixed price bids (FFP) or for materials and labor on a cost plus fee basis.

Conclusion: Summary of approach and reasons why this solution is preferred.

References: Specifications for equipment, resumes of key personnel, relevant corporate information, copies of certification documentation, etc.

Proposals are usually read by a team with qualifications in management, technology, and finance. The following questions are typical of those used to evaluate proposals:

- ✓ Does the proposal demonstrate a clear understanding of the problem?
- ✓ Will the solution work?
- ✓ Is the organization qualified to do the work? Has it done similar work? What is the quality of the references?
- ✓ Are the people on the team experienced?
- ✓ Does the proposal address all the RFP requirements?
- ✓ Are the costs justified and/or competitive?
- ✓ What is our involvement in the project? How will we work with this organization?
- ✓ Is there clear reporting and a good approach to solving problems?

There are some key factors to remember when writing proposals:

Time Is of the Essence!

A competitive proposal has a strict deadline. If you miss that deadline, the proposal is disqualified and all the effort put into it wasted. Schedule planning is essential along with good teamwork and parallel development of proposal sections.

Establish a Theme or Themes along with Your Technical Approach

Developing the theme of your proposal at the beginning will allow you to weave that theme through your document. For example, your theme might be that your design is preferable to all other designs, or that you are solving a problem that nobody else has thought about.

Use a Convincing Style

- Use active voice and strong words: "We are certain" instead of "We think," etc.

- Avoid passive verbs: Use "Our systems engineers can install the system" instead of "The system can be installed by our systems engineers."

- Summarize the "conclusion" at the beginning:
  "The XYZ billing system will provide ABC Telecom with a distinct competitive advantage. It costs less than other systems and has more functionality to provide flexible billing plans ."
  <u>Not</u>: "Our billing system costs less and has more functionality which provides flexible billing. Therefore ABC Telecom will have a competitive advantage by using the XYZ billing system."
  The reason for using the first version is that it emphasizes the impact statement instead of starting with a general observation. You want to get the reader's attention and clearly state why your proposal should be accepted.

Itemize and Document Cost Estimates

Estimating the amount of labor and materials for performing the work is almost an educated guess since every project is fraught with unknown risks. In order to minimize the risk, address any potential questions, such as:
- How will change orders be handled?
- How will the customer document the current environment?
- Who will provide user requirements?
- How will the system be accepted?

ALTERNATIVES

CRITERIA	Weight	Option 1 Rating	Score[1]	Option 2 Rating	Score[1]	Option 3 Rating	Score[1]
Criterion 1	1	4	4	3	3	3	3
Criterion 2	2	3	6	2	4	2	4
Criterion 3	3	1	3	3	9	1	3
Total	6	8	13	7	16	7	10

[1] Score = Rating × Weight

Invoice

A company needs to get paid for its goods and services; an invoice (also called a statement or bill) is used to request that payment. Although an invoice is not a long report, it is very important since it is used to notify a customer of payments due. An invoice reports on the work done for a given time period including deliverables, labor hours, materials, and other goods and services received. It also provides the customer with information about making the payment such as mailing address and due date. The invoice is short, but must be accurate or the customer will – justifiably – balk at making payment. And because the invoice is almost the last document received by the customer, an erroneous invoice can leave a lasting negative impression. Most engineers will not prepare an invoice, but will furnish information about their labor and other billable items.

Following is a typical invoice format:

Invoice

To: XYZ Agency

Attn: John Sullivan

Contract No: 1446

Invoice No: 1446-1

Project: CCX System Requirements

Deliverable: Preliminary Requirements Document

Labor: 1,252 labor hours × $86.00 per hour

Expenses (details attached): $1,633.00

TOTAL: $109,305.00

Please remit to:

ABC Corporation
1224 Technology Drive
Alexandria, VA 21442

Attn: Accounts Payable (1446-1)

Technical Reports:

Technical reports describe the progress or results of scientific or technical research and development. They are usually produced in response to a specific request or research need, and serve as a report of accountability to the funding organization.

Technical reports fall into one of two categories:

- Government-sponsored research reports
- Privately funded research reports

proposed contract changes. In these situations, technical staff members need to be aware that they will have to perform work under the conditions of the contract. The technical team needs to be flexible (or they will not get the work!), but also aware of the potential pitfalls that could cause them to exceed the budget for the project.

To write a contract, the legal team usually begins with its standard legal agreement and tailors it for a specific business deal. The legal team for the customer will then review the contract and may suggest modifications. During negotiations between the two companies, the sales and technical teams may be involved to further refine the details of the proposal, perhaps working on a final price acceptable to both sides. Throughout this process, the technical team needs to communicate their needs clearly to the legal team. For example, a customer's request for ownership of intellectual property rights for software may not be consistent with the technical team's strategy regarding software reuse.

As a technical professional, you may find yourself in the middle of an ethical dilemma. The management of your company may pressure you to rush through a proposal and contract, and to emphasize cost savings. This may be necessary for your company to win the contract. But in your heart and brain, you may know that the proposed solution will not accomplish the objectives and that your company will either end up delivering an inferior product or asking for modifications in the contract. In the past, engineers could respond by saying, "I did what I was told". But increasingly, the technical professional is being held accountable for the quality (or lack thereof) of the product. Let your conscience be your guide in making decisions on how loudly to object to certain approaches. And if public safety is involved, we urge you to avoid any compromises. You may find it interesting to research the causes for the Challenger disaster. This will give you insight into some of the ethical issues faced by engineers.

Good contracts reduce conflicts

Master Agreements

Some contracts are referred to as master agreements, intended to govern all transactions between two parties. For example, a master service agreement may specify that your company provide certain services according to different statements of work. You may be asked to write a statement of work (SOW) governed by a particular contract, so you will have to read the contract to find out the overarching terms and conditions of the work.

Make sure you understand what the contract means and don't be afraid to ask your legal team (if your employer has such a team or department) to explain unfamiliar terms. Even familiar terms may have varying interpretations when used in a legal document. For example, legally A or B means either A or B, not both. In mathematics, A or B means A or B or both (the inclusive or).

When you are writing an SOW for a client, make sure that you understand everything necessary to get the work done. Usually you will have to ask questions to get additional details; if you can't ask questions, entertaining, interesting, or standard. If your title is witty, it will get more notice than the standard note; if it is interesting, it will be more likely to be opened and read. Be cautious of spam filters. Titles such as "Important," "Critical information," or "Don't miss this" can easily trigger spam filters and your intended recipient may never see your email.

If a response is requested from the reader, it should be noted in the title or at the beginning of the email. Otherwise, the reader may give the note a quick scan and completely miss the requested action. If you wish to have a response quickly, state this in the subject line of the email. If, of course, you need an instantaneous answer, you should consider phoning as an alternate form of communication.

The body of the email should be short and to-the-point. Email is intended for quick communications; lengthy documents can be sent as attachments to the email so the reader can download and store them for later study. If you do attach something to the email, mention it in the body of the email so the reader doesn't overlook it. If the attachment is short, just copy it and put it into the body of the email; there is no need to make the recipient open it.

Be careful not to put too much into a single email. Because people read emails quickly, there is a tendency to overlook some or all of the contents of a lengthy note. If you want to make several points, clearly state this at the outset and enumerate the points. If you need to ensure that someone has received the email, ask them to respond.

Remember that using all capital letters makes writing more difficult to read and – on the Internet – it is considered to be shouting. Expressing emotion through excessive punctuation (!!!!) or symbols such as emoticons is commonly practiced in personal emails; professional emails can be witty but do not need additional touches.

Tip:

If you wish to have questions answered, you can state them clearly in the email. You can also note this in the SUBJECT line along with the number of questions. If you just have a quick question, say so – and you will probably stand a good chance of getting it answered. Following is an example of this approach:

. Subject: Quick question (Alternatively, subject could be "Important question" or "Three questions.")

Hi Rebecca,
Great news! I have scored two tickets for the concert tonight which starts at 8. Three questions:
     Can you make it?
     Dinner first?
     What time is good for you?
Hope to see you tonight!

After writing your email, take a few moments to review it before clicking the SEND button. Drafting an email and then reviewing it is a good practice if you are feeling unsure about sending it. Remember that you can't retrieve your email after you have sent it! (Some email systems permit the recall of an email, but the effectiveness of this feature depends on how quickly the recipient receives and opens the email).

Some things to check before sending:

- Do you have the right people on your distribution list? Have you included every-one? You don't want any hurt feelings!

- Look at the title. Have you stated the subject clearly and succinctly? Do you indicate the urgency or request of the email?

Review email carefully
before clicking SEND

- Review the body of the email. Have you checked the spelling? Is the email intent clear? Is the topic stated at the beginning? Have you enumerated any requests? Are deadlines, dates, and times clearly stated?

- C should be used carefully. If you include addresses in the cc field, a copy of the email and any attachments will go to each of those addresses. In fact, the only difference between that and the "to" section is that the recipient can see if they are the main addressee of someone who was copied and therefore may not have to act on the email. If someone who receives the email chooses to "reply," the response goes only to the sender of the original email. If, on the other hand, the recipient chooses "reply all," the response will go to the sender *and* all people copied on the original message.

- Bcc (blind courtesy copy) has several uses. If you list an address in the "bcc" field, that address will receive the original email and all attachments. However, the other people (the addressee and those on the cc list) will not see that address on their copies of the email, and will therefore not know that you have sent it to those on the "bcc" list. If anyone receiving the email chooses to "reply to all," the reply will not go to those on the "bcc" list. Other uses of "bcc" were addressed in Section 2.4.

Here are some examples showing different styles used for personal and professional emails. Please note that we have not included dates because email systems automatically insert them.

Personal email:

From: johncalder@acg.com
To: rtaylor@ucc.edu
Subject: Concert tonight?

Hi Rebecca,
Great news! I have scored two tickets for the concert tonight which starts at 8. Three questions:
> Can you make it?
> Dinner first?
> What time is good for you?
Hope to see you tonight!

Internal email:

From: johncalder@acg.com
To: ICS Group
Copy: carlanderson@acg.com
Subject: **Brown bag meeting tomorrow at 12 – RSVP Please**

      We will have another "Brown Bag" lunchtime meeting tomorrow in the 4th floor conference room at noon. Our own resident expert, Carl Anderson, will talk about "Advances in Network Security." Bring your own lunch; drinks and dessert will be provided. We look forward to another lively discussion, so bring your questions too!
      Please let me know if you are coming so we will have enough chairs and refreshments. Thanks!
      Hope to see you then. John

External email:

From: johncalder@acg.com
To: gwash@flyhi.com
Subject: **Facility tour questions**

Dear Mr. Washington:

My colleagues and I are excited about your offer to give us a tour of your facility. We are looking forward to learning more about your capabilities to design and manufacture our new equipment.
      •What are some possible dates?
      •How many employees may we invite on the tour?
Please feel free to call me at (212) 387-6221 at your convenience. I look forward to seeing you soon.

John Calder, Program Director
ACG Corporation

Responding to Emails:

Because email is so quick to send, it is also expected that a response will be sent quickly. If you are leaving the office for an extended period (e.g., travel or vacation) and will be unavailable to respond to your emails, set up an automatic response to notify senders that you are not receiving emails, dates of your absence, and alternate contact information.

Some responders leave a long trail of ping-pong messages that makes the latest message extremely long. Unless these messages are important, delete all but the last one or two from the chain. When someone has responded to your request, be sure that YOU respond with a quick thank-you note. Remember, we are all more likely to help someone who appreciates it!

Forwarding Emails:

Unless you are certain that the sender will not object, ask permission before you forward emails. By the same token, if you don't want to have your email forwarded, be sure to say that in the email. You should always be aware that your emails can be forwarded, so be careful what you put in an email. If you compose an email when you are upset or stressed, save it as a draft and re-read it again before you decide to send it. Remember too that you cannot retrieve most email messages!

Managing Your Email:

As a professional, you will probably spend considerable time reading and responding to emails. Be sure to check your emails often since many will be time-sensitive. If an email is notifying you about a meeting or event, be sure to respond quickly to the sender to let him or her know it you can attend.

Set up filters for incoming mail that will automatically shunt emails to proper inboxes: project, interest category, personal, or general. All email programs have a system for filing emails. Just as you should retain relevant documents in a paper file, keep your emails separated by project, interest, or sender until you are certain you will not need them.

Because you may need to have a record of responses to emails (including commitments to dates, meetings, etc.), you should not delete copies of your "Sent" messages.

We conclude with a warning. Although it is easy and free to set up personal email accounts (e.g., Yahoo), many technical professionals prefer the convenience of a single account – the one provided by their employer. Legal actions have clearly shown that employers have the right to monitor any computer use that involves company resources (e.g., work time, company computers, company network accounts). So think twice before you use your company email account for personal business. Ignoring this advice could result in anything from embarrassment to the need to update your resume. This leads us to the upcoming sections on the effective use of communications to help secure a new job.

3.5 *Memos*

A memorandum, or memo, is an internal communication written for several reasons, such as to:

- give timely information to others for decision-making, discussion, or problem resolution
- create a real-time record of a situation or information that may be needed in the future (also called a "memo for the record")

Because most memos are internal, the format and tone are more casual than those of a business letter. They also should be very direct and short. A memo generally has the following format:

	INSTITUTION OR COMPANY NAME
	MEMORANDUM
TO:	List of Recipients
FROM:	Author Names and Initials
SUBJECT:	Topic of Discussion
DATE:	Calendar Date
CC:	Names of those receiving courtesy copies
	Attachments, if any

- Statement of the purpose of the memo
- Discussion of the topic
- Requests or required actions

This is not the only format for a memorandum; the standard format established by an organization may have some variations. For example, some organizations do not wish to have blind courtesy copies (see Section 2.4) while others find it perfectly acceptable. For example, an manager may send a blind courtesy copy to her supervisor to inform him about project issues, but would like her team to continue working on a resolution without feeling the additional pressure of management oversight.

A "memo for the record" is often written to document evidence of a decision, work effort, or situation after its occurrence. An example is a memo written by a team leader to confirm a decision the project manager has made during a telephone conversation.

Some tips for writing good memos:

- Leave room for all senders (listed in the "From" section) to initial the finished memo following each member's name.
- Be concise; the memo should be brief, a page of written text at most.
- State the point of the memo in the first paragraph; managers may not want to read the entire memo.
- Justify the memo – why are you writing it?
- Avoid contractions and abbreviations unless they are very common and will be easily understood by all readers. If using an acronym that is not known by everyone on the distribution list, define it the first time it is used.
- Do not assume that all the readers have a technical background

Rules for Attachments:

- Refer to all attachments in the body of the memo in order and make sure that they are ordered correctly and don't get separated from the memo.
- Clearly label all attachments and do not include any that are not mentioned in the memo.
- Make sure your company name is on all attachments.

- Include specific directives on attachments, such as "For Internal Use Only."

Don't make them dig for key points

3.6 Business Letters

Although email is increasingly used for both internal and external communications, letters are used for official communications outside your organization. Business communications rely on letters because they are more formal than email. Communication by letter is more permanent and exact than in a telephone conversation or meetings, where information can easily be misunderstood or forgotten. So, even though it takes more time to write a letter than to pick up the telephone, the letter may be the preferred means of communication since it provides a clear statement of the communication and a record for later reference.

Letters are used to make introductions or inquiries, to convey information and to request information or cooperation. Writing a good letter is an essential skill for an engineer or scientist, especially for managers and executives.

Types of Letters

Technical professionals use various types of letters for different purposes. Some of these letters are:

- Cover letters for resumes and other documents (also known as transmittal letters)
- Thank you letter
- Job acceptance letter
- Reference letter
- Inquiry letter
- Information letter

Because resume cover letters are so important for marketing yourself, they are discussed in greater detail in Section 3.8. They were also discussed in Section 2.6 in the context of student communications.

Letter Format

If your organization has a business letter format of its own, you should use it. Otherwise, use the standard business letter format, which usually has a business letterhead at the top of the page and is either full-block formatted, with every line starting at the left margin, or modified-block formatted, with the heading and the closing aligned at the center of the page. Because the letter is a formal communication, avoid using abbreviations.

Business letters consist of the following parts (samples are given following the discussion):

Heading

Date

Recipient Address

Salutation

Body (to include the following four items):

1. Reason for the letter
2. Statement of facts
3. Request or follow-up
4. Contact information

Conclusion

Notations

Heading

If you are using letterhead stationery, insert the date two lines below the bottom of the letterhead, aligned at the left margin. Spell out the name of the month in full.

If you are not using letterhead stationery, begin with your complete address 1 to 2 inches from the top of the page. Spell out words in addresses such as "Street," "North,, etc. The state name may be abbreviated using the two-letter, all-capitals designated by the United States Postal Service. Include the date aligned at left with the address.

Recipient's Address

Two to four lines below the date, place the following items:

- The recipient's title (such as Mr., Ms., or Dr.) and full name. Address a woman who does not have a professional title as Ms.

Example 1: Transmittal Letter

Heading

ABC Corporation
3313 Technology Place
Seattle, WA 87778

Date

May 23, 2012

Recipient
Address

Meredith Brown
Program Manager
Federal Aviation Administration
1700 Crystal Drive
Arlington, VA 23312

Salutation

Dear Ms. Brown:

Body

ABC Corporation and NetEx are pleased to submit the enclosed proposal for your consideration in response to the Federal Aviation Agency's Request for Proposal 2442-A. Our approach for Air Traffic Control System Support relies on proven technology along with custom modules to meet the unique requirements of the FAA.

Conclusion

We look forward to hearing from you and welcome your questions. Please feel free to contact me at (934) 555-8996 ext. 54 or by email at staylor@abc.com.

Sincerely,

[Signature]

Sharon Taylor
Vice President, Marketing

Example 2 Thank-you Letter.

ABC Corporation
3313 Technology Place
Seattle, WA 87778

March 17, 2012

William Bristol
Vice President, Strategic Alliances
NetEx
989 Corporate Garden
Seattle WA 87797

Dear Mr. Bristol:

Thank you very much for taking time meeting with me to discuss a possible joint proposal to XYZ Corporation. Your suggestions were very helpful, and I would like to further explore with you the possibility of our two companies teaming up for this and future projects. As we discussed, our two companies could form a strategic alliance in a number of markets.

I hope to move forward on this as quickly as possible. I will review our discussion with Mr. Raines and call you in a few days to discuss the next steps. If you wish to reach me, please call me at (934) 555-8996 ext. 54 or by email at staylor@abc.com.

Sincerely,

[Signature]

Sharon Taylor
Vice President, Marketing

SUMMARY OF THE INVENTION The present invention seeks to provide a bagless vacuum cleaner which is more convenient for a user to manipulate. Accordingly, the present invention provides a bagless vacuum cleaner comprising a collecting chamber which is removable from a stowed position on a chassis of the vacuum cleaner, the collecting chamber comprising an inlet for receiving a dirt-laden airflow, an air outlet, a collection area for collecting, in use, dirt and dust which has been separated from the airflow and wherein part of the chamber wall in the region of the collection area is a closure member which is movable between a closed position in which the closure member seals the chamber and an open position in which dirt and dust can escape from the collection area, the chamber further comprising releasing means for releasing the closure member from the closed position, and wherein the releasing means are inhibited from releasing the closure member when the separator is in the stowed position.

DETAILED DESCRIPTION OF THE INVENTION Referring to FIGS. 1 to 3, a vacuum cleaner 10 has a main chassis 50 which supports dirt and dust separation apparatus 20. The lower part of the cleaner 10 comprises a cleaner head 22 for engaging with the floor surface. The cleaner head has a downwardly facing suction inlet and a brush bar is mounted in the mouth of the inlet for agitating the floor surface. The cleaner head is pivotably mounted to a motor housing 24 which houses the motor and fan of the cleaner. Support wheels 26 are mounted to the motor housing for supporting the cleaner and allowing movement across a floor surface. A spine of the chassis 50 extends upwardly from the motor housing 24 to provide support for the components of the cleaner. A cleaning wand 42 having a second dirty air inlet 43 is connected by way of a hose (not shown) to the chassis at the base of the spine 50. The wand 42 is releasable from the spine 50 so as to allow a user to carry out above-the-floor cleaning and cleaning in places which are inaccessible by the main cleaning head 22. When the wand is fixed to the spine 50, the wand 42 forms the handle of the cleaner and a handgrip 40 at the remote end of the wand 42 allows a user to maneuver the cleaner. These features of the cleaner are well known and have been well documented elsewhere and can be seen, for example, in cleaners which are manufactured by DYSON.TM., and thus will not be described in any further detail.

U.S. Patent   Mar. 6, 2007   Sheet 1 of 6   US 7,186,283 B2

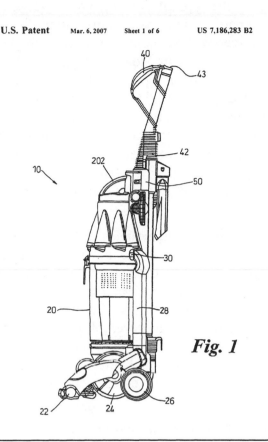

Fig. 1

The most important part of this document is the claims section which contains 26 claims (we have only reproduced the first two). You can see that this is written for and by members of the legal profession. Legal writing is different from technical report writing. As one example, note that all 149 words in claim #1 are in a single sentence! The writer must include all significant claims and not leave anything open to interpretation, even if this leads to very awkward sentence structure.

The "Background" section is easier to read. Although it is not specific to the claims of the patent, it does set the stage to understand why the inventors are claiming to have done something worthwhile.

The "Summary of the Invention" returns to long run-on sentences as it tries to describe the various claims in more general terms.

Depending on the type of position you have within a company, you may or may not participate in patent applications. Some engineers go through an entire career without contributing to a single patent. Others have their names associated with dozens of patents. You may wonder about remuneration for receiving a patent. There normally is none. If you work for a large company, you might get anything from a pat on the back to a token award (one large think tank purchased a textbook for any employee who wrote a patent). Indeed, contributing to the writing of the patent application is normally considered part of your job. The company usually owns the patent (unless you are prepared to prove that no company resources were used: e.g., you did this work on weekends in your garage).

Universities have varying policies dealing with faculty patents. These range from allowing faculty to patent inventions in their own names to assigning the patent to the university. It's harder for a faculty member to prove that an invention was conceived outside of assigned employment time since a faculty member's work schedule is very flexible. Indeed, as more and more employers move toward flexible work schedules and telecommuting, it is becoming increasingly difficult to separate work time from personal time. It is becoming harder and harder to act as an independent inventor while working full time as a technical professional.

If you patent anything as an individual, you do not receive any payment unless you either follow through and manufacture and sell the invention, or you sell the patent to a company. So while patents might serve as a mark of achievement (don't forget to put any patents on your resume), they are simply an enabling tool that allow you to prevent others from using your inventions for profit.

3.8 Employment Letters

Technical professionals may write several types of employment-related letters including:

Marketing letters
- Target marketing letters such as resume cover letters

- Response letters to position posting or advertisement
- Networking letters
- Search firm letters
- Mass mailing

Follow-up letters
- Thank-you letters
- Interview letters
- Acceptance letters
- Removal from consideration letters

Resume Cover Letter

It should be noted that Section 2.6 of this text discusses resume cover letters as they apply to students seeking their first professional employment position (entry-level). There are a few differences between letters written by students and letters written by those already in the profession. This section gives advice for technical professionals seeking new positions. In some cases, similar information from Section 2.6 is repeated here.

The most common marketing letter is the resume cover letter, which you need to tailor for a particular opportunity or organization. If you have seen a job posting, you should customize your cover letter for the position, highlighting how you would fit into the organization. In some instances, you may write untargeted cover letters inquiring about the possibility of an opening in a particular organization, but these letters are not usually as successful as targeted ones.

The cover letter is critical because it essentially introduces you to a prospective employer and describes how your education, skills, and background match the employer's needs. It summarizes your resume and can further emphasize those skills that best suit the company's needs. The letter can also direct the reader to the most significant accomplishments listed in the resume.

Before you write the letter, research the employer and the position or field using company annual reports, web sites, and other resources found in your library. The more you know about the organization, the more effectively you can spotlight the appropriate aspects of your background. Since a cover letter is the first thing a prospective employer sees when he or she reads your application, be sure that your letter is visually appealing and well written. A favorable first impression will make the employer want to read your resume and invite you for an interview. Make sure your cover letter is absolutely perfect with no misspellings or grammatical errors. Have someone else read it carefully to be certain that you have not made any spelling or grammatical mistakes.

Cover Letter Format

A cover letter conforms to business letter guidelines. You should use the same high quality bond paper that you use for your resume and mail it in a matching envelope. A separate cover letter should be written and produced for each opportunity and should be limited to one page.

The parts of a cover letter are as follows:

- Your address
- Date
- Employer's address
- Salutation (Dear Dr./Ms./Mr. with specific name)
- Text (three or four paragraphs)
- Complimentary closing (Sincerely, Yours truly,)
- Signature
- Typed name
- Enclosure (attachments)

Your cover letter will be much more effective if you address it to a specific individual by name, not just by his or her title. You can refer to business directories in the library or web-sites or you can call the organization to obtain the name, correct spelling, and title of the person and department to whom you should send your resume. If the receptionist cannot give out this information, ask to talk with someone in the Human Resources or Personnel Department to find out if they can give you the information or will forward your resume to the correct department. If you are unable to obtain a contact name, it is better to write something like

"Dear Director of Information Technology" in the salutation, rather than "Dear Sir/Madam."

Cover Letter Format

Heading/Salutation

Type your address, spelling out street addresses completely, centering this at the top of the page. Leave one blank line and insert the date, again spelled out completely. Leave a blank line and type the name of the organization and its address. Leave another line before typing the salutation: "Dear Dr./Mr./Ms. [*Specific Name*]:.

Opening Paragraph

State your reason for writing. Identify the position or type of work for which you are applying. Include a sentence about why the position interests you, and/or how you learned of the opening. If you are inquiring about an advertised position, indicate how you heard about it. For example, "I am responding to your recent advertisement in the *Washington Post* for the position of software engineer." If someone known to the reader referred you, state this. Don't forget to mention your attached resume.

Middle Paragraph(s)

Give the reasons why you want to work for this employer or in this field. You should include any information from your research on the employer or field and relate your background and qualifications. Emphasize or expand those parts of your resume that are most relevant to the position.

Explain what aspects of the industry, company, and position you find most interesting and how they fit with your career goals. As appropriate, refer to informational interviews and any research you have conducted.

Don't forget to mention your attached resume if you did not do so in the opening paragraph. As an experienced professional, your work experience will be most relevant and interesting to the hiring organization. You should state specific examples of your background that illustrate your skills and abilities to differentiate yourself from other candidates. Be sure to state your accomplishments using the active voice, with specifics whenever possible.

Last Paragraph

State the expected next steps and express your interest in discussing the position. You should ask to be contacted and provide your phone and email information. If appropriate, you can state that you will contact the person; for example, "I will call you next week to follow up."

Closing

Close your letter with a complimentary phrase such as "Sincerely," "Sincerely yours," or "Yours truly" two lines below the last paragraph of your letter. Leave four lines for your signature and then type your name.   On the line below, type "Resume Attached" or "Attachment".

You may be sending most of your resumes via email, but they should still be prefaced by a targeted cover letter. For an email cover letter, your address information is at the bottom of the email after your signature or typed name.

The various components of the resume cover letter are summarized in the following checklist

---

- Tell why you are sending a resume: in response to an ad, at someone's suggestion, or other motivation.
- Tell exactly how you learned about the position: a friend, website, or the reputation of the company in your industry.
- Tell why you would like to work in the organization and where.
- State aspects of your background that relate to the position.
- Show how you could benefit the company.
- Be sure your letter is well written (no mistakes!) and targeted to the company.
- Convince the reader that you are perfect for the position.
- Be sure the letter reflects your interests and personality.

Sample Resume Cover Letters

Email letter:

From: ejbauer
To: HR@Maelstromengineering.com
Subject: Project Engineer Position

Dear Susan Middleton, Director:

I am responding to your posting on monster.com for the position of Project Engineer. As you can see from my attached resume, I have over six years of experience in the construction field and have excellent client management skills. My B.S. degree from Virginia Tech in construction engineering gave me a solid education that provides me with the perfect background for the opening you describe.

Maelstrom Engineering sounds very interesting, and I would appreciate the opportunity to learn more about your goals and challenges. Although I currently live in Virginia, I can easily move to the Denver area. I hope to discuss this exciting opportunity with you. Please contact me at (703) 555-7778 or by email.

Thank you for your consideration,

Edward J. Bauer
3330 Pinecrest Drive
Alexandria, VA 22000

[Attached resume]

Regular mail letter:

1429 Pine Street
Philadelphia, PA 19887

March 17, 2012

Bruce C. Smith, Director of Software Development
Millennium Software
2000 Enterprise Drive
Wilmington, DE 20011

Dear Mr. Smith:

I have read your advertisement in the January 13 issue of *The Philadelphia Enquirer* for the position of Software Development Manager at your office in Philadelphia. I would welcome the opportunity to leverage my management skills to ensure a successful launch of your new product offerings.

As detailed in my attached resume, I have the required education for the position, including an MBA from the Wharton School. My professional experience includes five years of progressive advancement in software development for a variety of technical environments including Borland, Microsoft, C++, C#, .NET, Visual Studio, ASP, SQL, and Windows. In my current management position, I have introduced new software engineering guidelines that have resulted in on-time software delivery for all three major product lines and believe that I can achieve similar development improvements at Millennium Software.

I welcome the opportunity to discuss the position with you. Please contact me at (610) 555-9776 or via email at acmyer@yahoo.com. Thank you very much for your consideration for this exciting opportunity.

Sincerely,

Ann C. Myer
[Attached resume]

Networking Letters

If you are actively seeking employment, networking can be a very effective means of finding new opportunities or potential employers. If you have contacts in other organizations who can help you identify openings, a letter or email can pave the way to getting a face-to-face meeting. The goal of a networking letter is to get additional information or help, so you should make sure the recipient knows that you don't expect him/her to give you a job.

Thank -you Letters

When someone helps you in your job search, be sure to write a-thank you letter or email. After all, he or she spent time with you, so be sure to acknowledge your gratitude. In particular, a thank-you letter after a job interview is a chance to set yourself apart from other candidates and is another chance to acknowledge that you understand the position and would work well as part of the team. If you can include some information in your letter that you obtained from the interviewer, he/she will know that you were listening – an essential skill for a good employee.

The next page shows an example of a thank-you letter.

Thank-you letter:

1429 Pine Street
Philadelphia, PA 19887

May 12, 2012

Bruce C. Smith, Director of Software Development
Millennium Software
2000 Enterprise Drive
Wilmington, DE 20011

Dear Mr. Smith:

Thank you for giving me the opportunity to discuss the position of software development manager at Millennium Software with you yesterday. Not only did I appreciate learning more about the position; it also was very interesting to hear about Millennium Software's software applications challenges and strategy.

As you know, I am experienced in revamping existing software applications to ensure their continued viability and ease of maintenance, while conforming to IEEE software engineering guidelines. I am excited about applying my skills to help you achieve your goals on the software development team.

I enjoyed meeting with you and appreciate your consideration for the position. Please contact me at (610) 555-9776 or via email at acmyer@yahoo.com if you have any further questions.

Sincerely,

Ann C. Myer

Acceptance Letters

You've done it – you have the job! Now you need to formally accept it. Your job acceptance letter will become part of your personnel file, so make sure you acknowledge the position and start date. You probably will not state the salary and benefits since these are part of your formal offer letter. Congratulations!

3.9 Resumes

In Section 2.5 we discuss resumes in detail. The presentation of that section is intended for students. While there are many similarities between student resumes and those of professionals, there are enough differences for us to include this separate text section. It is designed so that you don't not need to cross-reference the section of the previous chapter.

A resume is a written summary of your education, experience, and accomplishments. The primary purpose of your resume is to get you to the next step in the hiring process – the interview. Your resume and cover letter are your marketing materials to a company, so they need to be professional, well-written, and on target for the company and position of interest.

Your application for a job normally goes through a human resources department. The division seeking a new hire writes a job description, which is used by the HR (Human Resources) department for its posting on the web, advertisement, or notice to a recruiter. The HR staff members – who are usually not technical themselves – filter resumes according the job description and send the best ones on to the hiring division. That division then decides which candidates they would like to interview and has HR set up the appointments.

Hiring Division → Job Description → HR → Posting
Resumes → HR → Qualified Resumes → Hiring Division

You can see that it is very important that you state your qualifications clearly in your resume. Make sure your resume contains all the key words or phrases that show you meet the criteria for the job as described in the posting. You cannot assume that the HR staff will be able to interpret your skills if you state them in a different manner from the job posting or advertisement. For example, if you are skilled in a particular computer language, list it by name on your resume rather than using a catchall phrase such as "4GL."

In addition to getting your foot in the door for an interview, a well-written resume:

- Provides basic information, including addresses and contact numbers
- Demonstrates your writing skills and professional standards
- Shows that you have the required education and skills to be considered for a position
- Gives information to discuss in the interview process
- Serves as a handout to people who can help you find the right position, such as those you meet at job fairs or conferences
- Is an attachment for job application forms, grant requests, and proposals
- Becomes part of your personnel records
- May be used by your company for further career development
- Can help you define and refine your job goals and direction
- Affirms your worth and gives you a boost of self-confidence – which really helps when you get that interview!

Many resumes are submitted electronically, but if you need to send a hard copy, be sure to print it on good quality paper. Don't forget to print copies to take with you on your interviews.

Resume Format

A resume is generally one or two pages long. A longer resume may discourage a thorough reading, so critical skills could be overlooked. Companies typically screen hundreds of resumes for a single position, so it is best to follow the standard layout:

Name (centered)
Address (centered
Phone and email (centered or balanced)

Job Objective: brief statement of your objective, tailored to the position if possible.

Qualifications: summary of your qualifications (optional)

Experience: description of your professional experience and/or assignments in reverse chronological order (i.e., most recent listed first). For each company, state the name of the company, the dates of employment and your title (s). Under each title and time period, describe your assignment, projects and accomplishments. Cite specific accomplishments such as percentages, dollar values, and time saved. If you haven't worked in at institution, project experience can be substituted.

Education: listing of the institutions attended, degrees conferred with dates and grade point averages (optional). Include any other professional training courses and certificates.

Accomplishments: statement of honors, awards, leadership activities, and professional memberships.

For recent graduates, the most recent and relevant aspect of the candidate's resume is his or her education. For an experienced professional, however, the description of work experience comes before the education section. Because it is more recent, it will be the focus of the job interview, so it should be detailed first. You do not have to include all your accomplishments in your resume. The best ones to include will be relevant to the hiring organization, will demonstrate the skills that the organization wants, and will highlight the areas of your interest.

A curriculum vitae (C.V.) is a very detailed professional history used in academia, grant proposals, and consulting to show all of a person's experience. In consulting, it is used to determine if a person is qualified for a particular assignment. It's always a good idea to have a C.V. of your own so you don't forget the details of your assignments or projects over the years. You never know when that project a few years ago might be useful in a future job application.

Some tips for good resumes:

- Use an easily-read font. The most common are New Times Roman and Arial. Times Roman is a serif typeface, meaning many letters and characters have small lines added. It is used by some newspapers. Arial is a sans-serif typeface. It contains no serifs.
- Use a size-12 and a 1-inch margin
- Avoid italics and all-capital letters as they are difficult to read
- Use active, strong verbs to stimulate interest
- Describe accomplishments, training, and skills in phrases, not sentences
- Use bold print or centering for emphasis (but not too often)

Electronic Resumes

Most companies use database systems to store, search, and retrieve resumes. Many companies and employment websites will ask you to submit your resume electronically. In this case, the computer will be the first reader of your resume and will scan for keywords and phrases. When you write your resume, include the keywords that are common in your profession as well as keywords that companies will look for such as:

Attributes: communications skills, ability to delegate or lead, results-oriented, problem solver, team player or leader

Industry skills: Systems, Software engineering, UNIX, Linux, java, database system, CAD, OO Programming, developer, network design, Cisco, mechanical systems, implementation

Adhering to standard formatting with common headings will ensure that the computer can easily scan your resume. White areas between the sections and bold headings will separate them into distinct parts. The computer can easily process Arial and New Times Roman fonts. Remember to be concise and specific, using only common acronyms and avoid using lines or graphics that are not easily processed.

Resume Writing Tips

- As you write your resume, remember to use strong verbs to emphasize action and provide results. For example, instead of, "Assigned to a design team working on improvements for aircraft safety." write, "As a member of the X718 design team, identified and incorporated three new features, improving safety of its exits by 32%."

- Don't be shy! Write about your responsibilities and accomplishments as if you were someone else. Don't oversell or exaggerate, but state the facts in a positive way. Remember that the reader really wants to learn about you and what you can do for his or her company.

- Your resume is not cast in concrete. Remember that you can tailor it for a specific position by highlighting certain aspects of your experience. For example, if the position is project manager, emphasize your experience in that area.

- Proofread and spell check every time you make a change.

- Let others read it before finalizing – get feedback, both overall impressions and editing.

- You may be surprised at the amount of time you will spend writing and re-writing your resume. Just remember the importance of a good resume – it is your main marketing material for yourself!

Following is an example of a resume that has been tailored for three different opportunities.

Opportunity A: The first opportunity is for a Director of Software Development of customer service applications for a company that wishes to improve quality, schedules, and profit:

JANE M. SMITH
5569 Shady Lane
Princeton, NJ 31110

Home: (271) 555-2223   e-mail: jsmith@gmail.com
Office: (433) 555-1465   Cell: (271) 555-3496

SUMMARY

Results-oriented executive experienced in project management and software development with a record of on-time, under-budget delivery. Recognized for leadership, creative problem solving, team building, and communication skills with solid background in software standards and packaging techniques. Accomplished in managing growth, fostering change, and developing staff in a dynamic, team-oriented environment.

PROFESSIONAL EXPERIENCE

TELPRO, Inc.
   2006 - present

Director, Communications Industry Services

Led the worldwide development, integration, and implementation of products and services for international telecommunications providers, including customer care and billing platforms.

- Developed product strategy that identified target markets and product fit, resulting in more focused sales and improved allocation of development resources.
- Conceived strategy to develop reusable, object-oriented software to improve scalability and reuse of software across two product lines, dramatically reducing costs.

- Established an integrated development group and introduced project management and software engineering standards to unite teams from four TELPRO divisions and improve software quality.
- Instituted processes to ensure development of complete, accurate plans and cost estimates for development and implementation efforts, resulting in project profit margins of 30%.

Cadence Management Systems   1999 - 2006
Director and Principal

As Director of Customer Care Systems, established a consulting practice focused on Customer Operations and Help Desk Systems
- Developed a strategy to help companies differentiate their services through outstanding customer service
- Implemented leading-edge solutions based on networked systems and various platforms including Oracle and SQL.

As a manager and consultant for CMS, managed development and delivery of consulting engagements, ensuring the attainment of group financial results, project schedules, and deliverable quality.

- Managed the development and installation of world-wide networks for a major New York financial services company.
- Developed a management strategy for a leading fast-food corporation, introducing leading-edge workgroup productivity applications and decision support systems.
- Created a strategy for directly linking Federal Mortgage Company's network-based mortgage offerings system with the Federal Reserve Bank.
- Formed and led network services practice for large installations in the U.S. and overseas.

Software Systems, Inc.   1991 - 1999

Project Manager and Software Engineer
As project manager, led successful software development projects through their complete life cycle, delivering on schedule and within budget.

- Produced functional requirements for the Customer Service System that was developed for the Internal Revenue Service.
- As Acceptance Test Director, assured the quality of all software produced for an intelligence agency by SSI.
- Introduced the use of IEEE software engineering standards and processes.
- Implemented leading-edge solutions based on networked systems and various platforms including Oracle and SQL.

PRIOR EXPERIENCE

Snowden Data Systems, Senior Systems Analyst
Universal Management Applied Research, Systems Analyst

EDUCATION

M.B.A., Virginia Polytechnic Institute and State University June 1988

M.A., Computer Science, Virginia Polytechnic Institute and State University   June 1987
B.S., Mathematics, *Magna cum Laude*, University of Virginia June 1985

Notice that this resume highlights Jane's most relevant experience for this opportunity and demonstrates that she has delivered the results that the company desires: higher quality and lower costs within schedule constraints.

Opportunity B:
Opportunity B is for a VP of Network Solutions. This is a new position in a consulting company, and Jane rewrites her resume to highlight her relevant experience:

JANE M. SMITH
5569 Shady Lane
Princeton, NJ 31110

Home: (271) 555-2223   e-mail: jsmith@gmail.com
Office: (433) 555-1465   Cell: (271) 555-3496

SUMMARY

Results-oriented executive experienced in consulting, project management, and network solutions with a record of on-time, under-budget delivery. Recognized for leadership, creative problem solving, team building, and communication skills with solid background in business and networks. Accomplished in establishing practice areas, managing implementation, developing successful marketing plans, and developing staff in a dynamic, team-oriented environment.

PROFESSIONAL EXPERIENCE

TELPRO, Inc.   2006 - present

Director, Communications Industry Services

Led the worldwide development, integration, and implementation of networked products and services for international telecommunications providers, including customer care and billing platforms.

- Developed product strategy that identified target markets and product fit, resulting in more focused sales and improved allocation of development resources.
- Directed marketing efforts for competitive bids, resulting in a four-fold growth each year.
- Established an integrated development group and introduced project management and software engineering standards to unite teams from four TELPRO divisions and improve software quality.

- Conceived strategy to develop reusable, object-oriented software to improve scalability and reuse of software across two product lines, dramatically reducing costs.
- Instituted processes to ensure development of complete and accurate plans and cost estimates for development and implementation efforts, resulting in project profit margins of 30%.

Cadence Management Systems   1999 - 2006
Director and Principal

As Director of Customer Care Systems, established a consulting practice focused on Customer Operations and Help Desk Systems

- Implemented leading-edge solutions based on networked systems and various platforms including Oracle and SQL.
- Developed a strategy to help companies differentiate their services through outstanding customer service.

As a manager and consultant for CMS, managed the development and delivery of consulting engagements, with responsibility for group financial results, project schedules, and deliverable quality.

- Managed the development and installation of world-wide networks for a major New York financial services company.
- Formed and led network services practice for large installations in the U.S. and overseas.
- Developed a management strategy for a leading fast-food corporation, incorporating leading-edge workgroup productivity applications and decision support systems.
- Created a strategy for directly linking Federal Mortgage Company's network-based mortgage offerings system with the Federal Reserve Bank.

Software Systems, Inc.   1991 - 1999
Project Manager and Software Engineer

As project manager, led successful software development projects through their complete life cycle, delivering on schedule and within budget.

- Produced functional requirements for the Customer Service System which was developed for the Internal Revenue Service
- As Acceptance Test Director, assured the quality of all software produced for an intelligence agency by SSI
- Introduced the use of IEEE software engineering standards and processes.
- Implemented leading-edge solutions based on networked systems and various platforms including Oracle and SQL.

PRIOR EXPERIENCE

Snowden Data Systems, Senior Systems Analyst   1989-1991
Universal Management Applied Research, Systems Analyst 1988-1989

EDUCATION

M.B.A., Virginia Polytechnic Institute and State University   June 1988
M.A., Computer Science, Virginia Polytechnic Institute and State University June 1987
B.S., Mathematics, Magna cum Laude; University of Virginia   June 1985

This version of Jane's resume shows that she has successfully established and managed groups (which requires marketing and team-building skills) and that she has experience delivering networked solutions. It should be noted that the experience for each company does not have to be in exact reverse chronological order. If you wish to highlight a particular aspect or project while at a company, you can place it at the top of the list.

3.10 Exercises

1. A draft contract states: "The software will be accepted once the agency is satisfied that it meets all requirements."

Is this an acceptable statement of acceptance? Why or why not? If not, rewrite the sentence so that it is acceptable.

2. Rewrite the following proposal paragraph:

It can be argued that the network may not have enough bandwidth. It is thought by the company that sufficient bandwidth exists within the network due to the usage model we developed. If the client does not like the network, it will be modified. It is an efficient network and low cost.

3. As the Vice President of Marketing, you have just completed a proposal for your company, ABC Inc., to provide an outsourced help desk in response to a government RFP. There are four people on the evaluation committee, Mr. Jones, Ms. Washington, Dr. Harris, and Ms. Perlstein. Write a cover letter for the proposal.

4. The Telcell Company is growing fast and now needs a new billing system for its cellular telephone service. Some of the requirements for the new system include:
> Carryover minutes
> Free calling on evenings and weekends
> Billing for text messages
> Flexible billing dates
> (any others you can think of)

Write a Request for Proposals.

5. Find an RFP on the Internet and critique it in writing.

6. You are in charge of a team that is scheduled to launch a software release at the end of the year. You have held numerous meetings with representatives of the internal Information Technology group (IT) to

inform them of your requirements for verification testing during September. It is the beginning of August, and IT has just told you that it cannot support your test due to an offsite training session.

a. Write a memo to your manager informing her of this situation and asking for her help in resolving the problem.

b. Write a memo for the record about your preparation for the test and the current situation along with recommendations for resolution.

7. You are part of a team that has discovered a new way to clean runoff water from farmland, removing excess fertilizer, herbicides, and waste runoff. How do you apply for a patent? What are your concerns? Who would be interested in your invention?

8. Write an email inviting your team to a project status meeting. Each person on the team will present a status report that you need to review before the meeting.

9. Write an email to tell your colleagues about a luncheon for a departing member of the staff.

10. Write an email asking a network equipment vendor to commit to a delivery date.

11. It is Friday morning and you have just found out that you are in charge of writing a proposal that is due in one week. You know that you will not be able to meet the deadline unless you and your team put in at least eight hours of work before Monday. You need to write an email informing your team about proposal work scheduled for the coming weekend. What details should you include in the email? What is the best way to word it?

12. You are a project manager for the CMIS system, which will be built using SCP software. You have just been notified by the supplier of the software that there will be a two-week delay in its latest release. Write an email to the supplier of the SCP software expressing your concern about the delay. Now write the corresponding letter to the supplier.

13. Write a cover letter for the following job posting which you found on monster.com for the WLN Company. The job closely matches your skills and experience:

Software Engineer

We are seeking a team-focused individual to support the wireless networking systems group in creating user interfaces in the GUI development segment of our product design group. This person will conceive and develop a wide range of Windows GUIs, tools, and applications for use in wireless network system set-up, operation, and administration.

Principal Functions:
• Conceive, design, modify, and implement Windows http client and Linux http server applications
• Design, modify, and implement software tools for configuring and testing existing and advanced applications
• Support advanced product development with new user concepts and GUIs
• Modify and adapt existing Windows interfaces for client-specific applications

Job Skills & Experience:
• Proficiency in Javascript and complex HTML user interface
• Server side CGI programming
• Experience in developing applications in PHP and Javascript
• Proficiency in C, C++, PERL, and PHP
• Familiar with one or more website development tools, such as Macromedia/Adobe Dreamweaver.

Qualifications:
• B.S./B.A. degree in computer science, math or physics
• 2-3+ years experience developing or supporting revenue products and/or systems
• Basic understanding of networking
• Familiar with Linux and Windows operating systems

• Ability to work independently, identify technical issues, and propose convincing solution.

Interested and qualified candidates should email their resume in MS Word format to:
Hank Jenkins
hankj@wln.com

14. Write your resume if you have not already done so. Ask someone to review it and incorporate comments. Can you determine any weaknesses? Do you have any ideas to strengthen your resume?

15. Rewrite the following resume for an entry-level engineering position:

Alexander T. Stevenson
639 Colfax Avenue
Columbus, OH
(313) 234-3051
ats@eot.edu

Experience

Assistant mechanic at Washington Auto Service, Lapeer, MI, summer '04, changed oil, etc.

On the team at Eastern Ohio Tech that won the ZOOM stock car award, sponsored by the Automobile Engineers Association. Fall 05 – Spring 06

Summer Intern, General Motors summer 2006, Detroit, MI, advanced design project using mechanical design, CAD/CAM composites design, worked on project team.

Student Co-op, Fall 2006, General Motors, statistical analysis of prototype

Education

B.S., Mechanical Engineering, Automobile Design concentration, Eastern Ohio Tech, Columbus, Ohio, June 2007

Honors and Activities

Intramural basketball 2003 – 2006
Resident Assistant 2004 – 2007
Paul Raymond Engineering Award 2006
Tau Beta Pi Engineering Honor Society, inducted 2005

Related Coursework
Calculus, physics, thermodynamics, statistics, circuits, fluid mechanics, design, automotive engines, etc.

Skills

CAD/CAM, Word, Excel, AutoCAD, MathCAD, C++

---

16. Make a list of your professional accomplishments. Now write a concise accomplishment statement for each one of them.

17. Which of the accomplishments in the prior exercise are best suited for your resume? Why or why not?

Chapter 4
Beginning a Communications Project

4.0 Introduction

Now that we have reviewed many types of writing required of engineering students and professionals working in technology, and your have heard about the importance of good communications, we are ready to start developing and improving those communications skills. This chapter discusses important topics for beginning any communications project:

- Identifying and understanding your target audience

- Making sure the audience will be interested in your writing or presentation

- Methods for gathering information

- Various team approaches to communications projects

Plan your project

After determining the interests and needs of your audience, deciding how to gather information, and determining how you will work with your team, you will understand the scope of your project and what needs to be done. You could begin your project before thinking about these issues, but without understanding whom you are addressing and what you are going to say, you will not fully understand the writing demands of your project. Any time spent at the beginning of the project which furthers your understanding will reduce unproductive time and reward you with a successful outcome.

4.1 Assess the Audience

Before you enter the first word of your document or presentation into your computer, you need to decide who will be reading your paper or listening to your talk. It is most important that you know both your topic and your audience so that you can impart your knowledge in a way that will be easily understood by others. In order for you to get someone else excited about reading or listening to your words, you must first be excited about writing them. The topic must be something you know and care about. In addition, the structure of your writing and presentations, your writing and speaking skills, the attitude you project both in print and in person all affect whether or not your audience will want to listen to or read your words.

In addition, the more detailed the information, the smaller the potential audience because of more focused interests and perhaps more required background knowledge. If you are writing or speaking about the Brooklyn Bridge, for example, many people will be interested to know about its history; fewer will care about the technique used for sinking the pylons into the East River; fewer still will want to read the calculations for determining the tensile strength of the cables used in this suspension bridge.

There are also times when certain people will be required to read your work or listen to your presentations: teachers, colleagues, employers, clients, customers, managers, maintenance staff, users, just to name a few. Your words and methods of presentation must differ according to the audience and intended use for your final product.

As a student, you would write or give a presentation one way if you are communicating with other students and another way if your writing is intended for your professor. As a professional engineer, your writing or speaking would be far different if intended for other engineers in your company as compared with presentations for the general public. Even within these groups there are additional levels of refinement. If you are writing a paper or giving a talk about your work on a project, and your intended audience is your fellow engineers, you would structure the presentation in a particular way. If, on the other hand, the readers or listeners were all part of your design group, they would already know a lot more about the subject than if they were from outside the group with little knowledge about your project. In this case, you would structure the presentation differently.

MARTIN RODEN'S PERSONAL EXPERIENCE

While teaching a technical writing class at the California State University, Los Angeles, I asked the students to write a very short paper about Ohm's Law. Ohm's law associates current, voltage and resistance and is related to Newton's second law, $F=mA$. Ohm's Law states that the resulting action (movement of an object or current in a wire) is proportional to the applied parameter (force or voltage) and inversely proportional to the property of the material that fights against the action (mass or resistance).

Members of the class immediately started doing one of two things. Some of the students started madly entering words into the computer as if they were taking a speed typing test. Others went to a search engine like Google to find information about Ohm's Law, which I hoped they were planning to use as background material and not to "cut and paste".

After several minutes of this, I stopped the class, strongly rebuked them (in a friendly way) stating, "Why didn't any of you ask me whom you are writing this for?" A textbook section on Ohm's Law would be far different from a letter to a non-engineer friend describing the technical concept. I then refined the assignment by saying the paper should be directed to a recent high school graduate preparing to enter a college engineering program. Of course, they produced far different papers from those they had initially started.

Approaches for Meeting the Needs of Your Audience

Once you have determined and understood the target audience, you need to give some thought to the approach you will use to develop your paper or talk. Using the Ohm's Law example in the "Personal Experience" box above, you will find that there are several different approaches you might use.

Let's call the first one *The Google Approach*. In this approach, you go to a web search engine and try to find relevant web sites. Then you take notes from these sources and structure your presentation from these notes using your own words. (Make sure to give appropriate credit to the source—See Section 6.6 on Plagiarism). In addition to the obvious risk of plagiarism, some of the potential pitfalls of this approach are:

- Your paper could end up looking like it was written by a committee with different styles and maybe even different notation formats.

- You learn little from this experience since you are essentially just summarizing writing done by others.

- You do not add your personal touch to the project, nor are you putting your personal touch on this project.

Suppose you used this approach to write the paper described in the "Personal Experience" box above. The start of your paper might look like this:

Georg Simon Ohm was a German physicist born in Erlangen, Bavaria, on March 16, 1787. As a high school teacher, Ohm started his research with the recently invented <u>electrochemical cell</u>, invented by Italian Count Alessandro <u>Volta</u>.

We'll call the second approach *The Clinical Approach*. This is the most straightforward, approach; it may also be called, "Just the Facts." This is the approach that textbook authors use in writing their books. Students <u>must</u> read textbooks as part of completing a course, so many textbook writers don't worry about making their writing interesting or fun to read (We hope this text is an exception!). The material doesn't have to be interesting: it must only be clear and accurate. Of course, there are some instances when the straightforward approach is necessary, especially when you don't know the background of your audience members. If you use The Clinical Approach for the Ohm's Law paper, the start of your paper might look like this:

> *Ohm's Law is fundamental to Electrical Engineering. It describes a relationship among current, voltage and resistance in an electrical circuit.*

We shall call the final approach, *The Audience-Geared Approach*. If you carefully assess your intended audience and are able to decide what will hold its attention, you can write an audience-geared paper or give an effective presentation. Getting the audience's attention in the first sentence is of paramount importance. You hope the audience will **want** to read or listen to the remainder of the presentation.

If you use the audience-geared approach for the Ohm's Law paper (remember it is meant for recent high school graduates), the start of your paper might look like this:

> *Did you ever wonder why the water coming from a faucet in a sink changes in volume as you turn the knob or move the lever?*

Many engineers and scientists are introverts; many are also perfectionists. Although they have lots to say, they can be afraid of ridicule and/or making a mistake. So they stay on the sidelines out of fear, perhaps criticizing those they think are more articulate, but have much less to convey. They know that engineers and scientists can be a tough audience; they can usually spot an error or weakness right away. Engineers sometimes hate to write or speak out of fear that no one will want to read or hear their presentation or will find fault with it. Sometimes they even justify their lack of communication skills with those reasons.

That attitude is self-defeating. Not only do engineers and scientists have something to say, once they have the right communications skills, they can really leverage their knowledge to inform, convince, and sell others on their ideas. After many years of study, research, and/or work experience, you become expert in your field. Your experiences, results, opinions, and accomplishments certainly do matter and are interesting to the right audience. When you put your knowledge into well-chosen words and create a document that is a pleasure to read or hear about, your audience will respond with their attention and more.

The "bottom line" in all of this discussion is that knowing your intended audience is critical to the acceptance of anything you create. You could be an excellent writer or speaker, yet if your projects don't address the audience, they will not be well-received. So take the extra time to clarify this issue before beginning to work on your communications project.

4.2 Getting – and Keeping – Your Audience's Attention

You have written a document – how do you make sure that your audience wants to read it? You have invited people to your presentation – how do you make sure that they want to come, and if they do, how do you make sure that you keep their attention?

- Make sure that your title is accurate and relevant to your document or presentation. It is the most important part of your work. Have you ever read an article that doesn't cover the topic stated in the title or makes you work very hard to find it? It is most annoying! Since your title will be the first and best way to get attention from your audience, spend some time thinking about it and ask others about it. Sometimes you can add detail in a description of the document or presentation that will interest people.

- Put some intriguing information into the cover letter of the document or into the invitation or notice for the presentation. State the problem that the document or presentation solves or involve the reader or listener in the issue. A question format works well to raise interest. For example, "Ever wonder how independent computer labs test new products?"

In your writing, try to envision your reader and talk to him or her. Make sure that you thoroughly cover the basics (e.g., spell out acronyms) at the outset so you don't lose your reader at the beginning. If you are making assumptions about the reader's knowledge, state them up front. If you have a purpose in your text, make sure that is clear to the reader.

For example:

> "The purpose of the attached document is to define the contingency operation plans for the ABC computer center. These plans are currently in draft form and under review by management and operational staff. All reviewers' comments, questions and corrections are due back to this office by May 23, 2008. If you have additional questions or concerns, please contact the Project Management Office at X4337.

Keeping Them Interested

- Add humor or pictures: Unless the setting is formal, begin your presentations and documents (where appropriate) with some humor and/or drawings that will keep people interested and involved. Vary your writing and slides to avoid repetition – people get bored after looking at the same thing for too long. Intersperse extremely detailed slides with others showing bulleted lists, photographs or drawings. If you are talking about numbers (a good way to put everyone to sleep), try to summarize, drawing attention to the important ones – and telling your audience why the numbers are truly important. If you can compare and contrast numbers and results to expected results from other occasions or periods, your audience will understand their importance more easily.

- Lighten up: People love to be entertained and you will be more relaxed and engaging if you inject humor into your writing and presentations (as appropriate). If you are presenting and something unplanned happens, regard it as an opportunity to enjoy a laugh with your audience – even at your own expense. They will love you for it! If you think you will be nervous, start your talk with a humorous story or comment; it will relax everyone and get your talk off to a good start

- Start out strong: Don't keep your audience guessing with poorly structured writing or presentation material. Get to the point right away; people are annoyed rather than intrigued when an author goes through a series of "Since..., therefore..." sentences. Tell your audience what you are going to tell them first: your results, your theory, your request. Follow that statement with the complete explanation, providing background, details, repercussions, etc. Always wrap up your paper or speech with a recap of your document or presentation in a concise conclusion.

- Keep them comfy during oral presentations: When giving a speech, check out the facilities before your presentation and make sure that there are enough chairs and that the temperature in the room is comfortable. Remember that adding your audience to the room will increase the temperature. Some presenters like to provide refreshments; others find that they distract from the talk. We think it can be a good idea if your meeting is informal and at time when your audience will likely need something to eat or drink.

- Speak up: It is important that your voice can be heard – make sure that everyone can hear you. Try to project your voice – without yelling – so that it is easily heard without a microphone. Your audience members will lose interest if they can not hear you.

Getting Them Involved

In oral presentations, maintain eye contact so listeners feel that they are in a conversation with you – the ideal way to get information across. Invite your audience to ask questions during or after your presentation. If you ask your audience questions, they can also feel involved in your talk and vested in your success. For example:
- Who has ever used this equipment?
- Who has ever heard of the ....?
- Are there any questions?

Barriers to Effective Communication

There are some factors that may get in the way of keeping your audience interested and involved, for both oral and written communications.

Oral Communications:

- Bias – Be aware that some may think you cannot possibly tell them anything. You can try to overcome this attitude with a humorous comment, but be sure to project self-confidence too.

- Language differences or accents – Try to minimize slang expressions and be sure to allow your audience to interrupt if they don't understand something you say.

- Noise or other environmental factors – Check out the room ahead of time

- Short attention span – Keep your talk lively; don't belabor points. If someone asks questions that are not relevant to your talk or are too detailed, offer to discuss them after the presentation. Keep your presentation moving and focused on the important points you need to make.

- Distractions – Eliminate distracting mannerisms such as nervous tapping, jangling coins, etc.

- Content– Keep your presentation concise and logical. Nothing loses your audience more quickly than rambling or overly-wordy slides.

- Style – Keep your style relaxed and natural. If you are tense, your audience focuses on you rather than on your message.

Written Communications:

- Clarity – Make sure your writing is clear and concise. Stay on the topic and don't add extraneous facts that sidetrack readers from the main issues.

- Format – Use a standard format that readers expect to see.

- Reading fatigue -- Use bullets and summary tables to break up text.

- Limited time - Include a summary for those who do not have time to read a long paper.

- Information overload – People have so much to read and absorb, especially with high-volume email. If you keep your emails short and snappy with good titles, you will be more likely to have them read.

Get feedback

Good writers and presenters are always trying to improve, so actively solicit feedback from readers or audience members. Invite readers to email you with comments or provide audience members with feedback forms. If you provide them at the beginning of your presentation, your audience may even be more attentive so they can fill in the form. When it comes to feedback about writing, email is often too labor intensive. Face to face meetings are the most efficient for the reviewer, so you might want to invite people to either meet with you or call you on the phone.

4.3 Gathering Information - The Literature Search

It is rare to compose a document by just sitting at the computer and typing (except for personal letters and emails). You will almost always be building on the work and writings or others. After all, you don't want to continuously reinvent the wheel.

But how do you find relevant information and documents? Here is where you must become a detective. Don't try to cut corners and save a lot of time. Any time spent tracking down resources should significantly reduce the time you need to spend generating information.

Fortunately, a variety of resources exist to help you in this search. We briefly examine:

- Library card catalog

- Abstract services - Engineering Index, Applied Science and Technology, Reader's Guide

- Individual publication annual indices

- Professional organizations – such as IEEE and Xplore

- Personal contacts

- Professors

- Internet

Library Card Catalog:

It takes several years for books to "catch up" with new technical developments. It takes less time for magazines and periodicals to be current. But these traditional resources form a good starting point to develop the necessary background information for your writing. Your college or company library is a wonderful resource, and you should start with the card catalog. This is usually available online. Start by looking for the specific topic. Then broaden your search as needed. For example, if you needed information on wireless internet, you could start by looking for that specific topic. If that fails, try just "wireless" or just "internet". You may even have to get creative and use search terms such as "portable computers." Sometimes if you find one good reference, you can develop others from it. A publication on the growth of wireless hot spots (perhaps focused on Starbucks) might give references at the end that can be used for more detailed research. Of course, none of these articles can look into the future, so you want to try to locate the most recent articles first.

Abstract Services:

There are millions and millions of articles published on virtually every technical subject. These articles represent current developments in the field and are typically less tutorial (i.e., not very pedagogical) than what you find in books. How can you hope to locate the relevant articles for a specific topic you are researching? The various abstract services assist you in this search.

The *Engineering Index* (and *Compendex*) has been a major resource for abstracts. This combined with *Applied Science and Technology* form the starting point for many literature searches. You look up topics in the index, and then are directed to individual articles related to those topics. Full text is often available online so you don't even have to locate the original article. These abstract services compile abstracts from many hundreds of periodicals, including the publications of all of the major technical societies. These are only available if your school or company library subscribes to them, so you may find some variation in which abstracts are available. Another useful abstract service is the *Reader's Guide to Periodical Literature*. This tends to be lighter in terms of the technical coverage. It presents abstracts from the more popular magazines such as *Popular Science, Popular Mechanics*, financial magazines, and *Consumer Reports*. While these articles will not be as technically complete as those in the professional organization publications, they can still serve as a good starting point to gain an overview of a topic. Some of the articles contain references and resource for further study.

If the topic you are researching is well-defined, you may know which publication is likely to have relevant articles. Must you then go to every individual issue of that publication and look at the table of contents? A better way is to go to the annual index of articles, which usually appears in the December or January issue of a monthly publication. If you start with the most recent annual index, you may find an article that contains references and can therefore lead you to other earlier articles.

We have one final bit of advice for using abstracts in your search. You can easily get carried away during your literature search. If you gather 20 or 30 papers before you begin your study, you will probably be overwhelmed, and most of your time will be spent trying to understand these articles. For example, some of the articles in the Transactions of specific IEEE groups can require weeks of study before you begin to understand what the author is saying (the articles are intended to document new knowledge and discoveries. They are not intended to be tutorial.). We suggest you limit your literature search to locate no more than six relevant articles.

Professional Organizations:

The professional organizations (ASCE, ASME, IEEE, ACM) publish a high percentage of the technical literature in existence. Many of these organizations have online resources. For example, IEEE Xplore is a major resource for locating articles in IEEE publications. Many of the online resources require that you subscribe, but your college or company library may already have a subscription to some of these. Check with the technical librarian for details.

Personal Contacts:

Perhaps one of your friends or relatives is a technical professional. Don't overlook the possibility of obtaining information from these people. Maybe you met someone at a technical presentation or a career fair (try to develop your own file of business cards). It takes very little time to contact someone, and at worst they will say they cannot help. If they are not familiar with the topic, they may be able to refer you to someone who is. Technical professionals usually love to talk about their work (unless it is confidential or embarrassing). They are proud of their accomplishments and enjoy sharing information.

Professors:

Professors enjoy working with students. That's why they have chosen teaching as a career. So don't hesitate to ask professors for help. Of course the professor who gave you the assignment may not want to "lead you by the hand" too much, but it doesn't hurt to ask for help. Just one note: be a bit cautious in approaching professors other than the one who gave you the assignment. You don't want to give the impression that you are trying to get help on something you should be doing on your own.

Internet:

The Internet is dynamic, massive, and all-encompassing. It is easy to get carried away and forget all of the other forms of gathering information. But you must bear in mind that there are two broad categories of contributors to the exploding list of web pages. The first category consists of those with a profit motive. The majority of websites, while they may be disguised to appear instructive, are really trying to sell something. So be cautious regarding the objectivity of a web entry. The second category is the relatively small percentage of web sites that have no profit motive. In many cases, these are university websites where a professor posts class notes or tutorials, or they may be personal blogs. Some of these non-profit websites may also be linked to a commercial website - for example, a semiconductor company may post "user notes" for a particular integrated circuit, and although these notes are keyed to that company's product, they still can yield a lot of general information.

The Internet contains lots of hybrids between the for-profit and non-profit world. Some excellent information-rich websites appear to be non-commercial until you focus on the numerous advertisements accompanying any article. Others are more devious in their commercial links. For example, be cautious of web search engines that seem to give prominent placement to links that pay the search company for the privilege of being listed early.

Be cautious regarding copyrights. A creative work (e.g., article) should be considered to be copyrighted even if the symbol © does not explicitly appear. An exception is those items in "public domain," which includes anything written by an employee of the U.S. government as part of their work. It's always best to assume written documents are copyrighted. Copyrighted material can be used for non-profit educational purposes, but this certainly does not cancel your ethical responsibility to credit the work. Be extra cautious of artwork and diagrams. For text, it is always better to state ideas in your own words, though a limited amount of quoting is acceptable.

The next section of this text highlights the major search engines that try to help you locate websites with relevant information for your research. The size of the internet is both an advantage and a disadvantage. The obvious advantage is that, if you have patience and detective skills, you can find almost anything online. The disadvantage is that you can be easily overwhelmed by literally millions of site referrals, most of which have nothing to do with what you are looking for. If, for example, you are searching for information about how a particular mechanical devices works, you may find yourself drawn to the web site of a small store (or even garage) halfway around the world trying to sell the device.

We close by emphasizing a point made earlier. Patience is important during the "gathering information" phase of a project. Time spent on thorough research can significantly reduce the amount of work you will have to do writing your paper.

4.4 Web Search Engines

There are over 100 million web sites available on the Internet. The number of indexed pages exceeds 20 billion (maybe 30 billion by the time you read this)! How can you possibly expect to find anything on the web unless someone gives a specific address (URL)? The answer lies in web search engines. These powerful tools permit the user to zero in on web pages addressing specific topics. Access to most web search engines is free since they are supported by advertisements. The advertisements can sometimes get quite aggressive since the financial survival of companies offering these services depend on their ability to sell advertising space.

Most search engines use "crawler-based technology," which uses spiders (robots) to crawl through the web. These continuously and automatically map web pages and find key words. One problem with search engines is the sheer volume of "hits." For example, suppose you wanted to learn about crawler-based technology. Entering that term into Google results in over 200,000 hits. Alternatively, if you enter "how crawler-based technology works," you get about 1.5 million hits. The additional words cause this search engine to include related sites that were not included in the original specific term (this is counter-intuitive, since adding descriptors to other types of searches such as abstract indexes reduces the number of matches). It takes skill and practice to learn to navigate through this and pick the most promising websites to

became a subsidiary of the Convex Group in 2003. It is also the exclusive online publisher for the Consumer *Guide and Mobil Travel Guides*. This has resulted in having many ads accompanying the articles, but the articles themselves appear to be free of commercial content and well written.

There are other more specialized web search engines. The only way to become proficient at locating web resources is through practice. We urge you to spend the necessary time to learn how to use these resources. In this information-based age, those who are skilled at accessing information will find it much easier to achieve success in their professions.

4.5 The Collaborative Process

Many of us were brainwashed during our early education about collaboration. We were taught that working in groups to do homework or take an exam was a form of cheating. When we were given any writing assignment, the implication was that we must create the document ourselves without any outside help.

You must work to counter this indoctrination. Universities realize the importance of teamwork and the collaborative process. Accrediting agencies insist on it as they review and evaluate university programs.

Industry has always understood this. Although some of your work time will be spent alone devising solutions to problems and performing analyses, most of your productive time will occur as a member of a team. One driving force behind collaboration is the fact that technology is becoming more and more complex. It is increasingly rare for one individual to know enough about a project and all of its aspects to be able to work alone. So you will have to function well in a team, and this includes collaborative writing during the documentation of technical work. There are numerous guides to team functioning, and ultimately, the way teams function will depend on the personality of your employer. Your colleagues probably have a particular way of functioning, and although you can suggest changes, you should start by adapting to their way of doing things. After all, it has probably evolved over many years.

In this text, we concentrate on hints for team writing – the manner in which a report is prepared with contributions from multiple authors. Between 60% and 80% of industry-prepared documents are created by teams rather than individuals.

Team leader

Although we all like to be democratic, someone must be in the position of arbitrating disagreements. This person also will normally have to make writing assignments to members of the group. The leader should be someone who is firm yet sensitive to individual skills and needs. Every member of the group has to feel that he or she is contributing to the overall result and that individual opinions are heard.

The leader also needs to keep team members aware of the progress and status of the writing project. This could involve periodic meetings and reviews. But it also could include setting up a computer tracking system, on which everyone's contribution is posted so others can see. In some cases there should even be a way for others to make editorial and substantive comments about particular sections.

The team leader should insure adequate interaction among team members. If each member is simply assigned to draft a section of the report, the result will not appear coherent. Such reports receive negative comments such as, "This looks like it was written by a committee."

Preliminary Steps

After the leader gives assignments to team members, each member should prepare a comprehensive outline of their section. The team should then review all of the outlines and discuss such details as graphics (charts, graphs, drawings).

Review by Team Leader

After each team member submits a draft, the team leader needs to spend a lot of time and effort reviewing the entire package. The leader should then assemble the entire team to discuss the report, make suggestions, and give follow-up assignments. Once the report has been assembled, the team leader must then either perform the final edit or find someone who is qualified and available.

How much of this team writing approach can be used while taking technical or writing classes in the university? You need to be cautious and become aware of each professor's rules. While we hope team writing is widely practiced at your university, we acknowledge that some professors would still consider it to be cheating.

Most accredited engineering programs require a senior design (often called "capstone design") course, and many of these permit (and perhaps encourage) team projects. This would be an ideal time to take the team approach to writing if you have that option.

4.6 Exercises

1. You have been assigned to give a presentation on trends in the telecommunications industry. What types of information are interesting to each of the following groups in your audience?

Engineers
Architects
Accountants
Marketing & Sales
Management

2. List some websites that could be useful in developing the presentation in the preceding exercise and the reasons why you chose each of them.

3. State, in writing, the Pythagorean theorem, which gives the relation among the three sides of a right triangle:

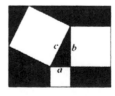

➤ Explain the theorem in writing as a teacher would to a class (What is it? How does it work? History?)

  - ➢ Discuss the theorem with your fellow students (Why do you like it? What can you do with it?)
  - ➢ "Sell" the theorem (what makes it special, useful, etc.)
  - ➢ What would change if it were not true?
  - ➢ Who should be interested in the theorem?

4. Suppose you need to give a presentation of your own choice:
  - ➢ State a topic of interest in your area of expertise
  - ➢ Formulate a title for a presentation
  - ➢ Who would be interested in your presentation?
  - ➢ How would you motivate people to attend the presentation?
  - ➢ What could you expect to gain from your presentation?

5. What are the pros and cons of searching for information in each of the following sources:
  - ➢ Library
  - ➢ Abstract services
  - ➢ Professional organizations
  - ➢ Personal contacts
  - ➢ Professors
  - ➢ Internet

Which source(s) do you expect to use the most? Why? Which publications are most useful to you? Which individuals?

6. What is the best way to record the information you find and keep track of your sources?

7. What are the general characteristics of a good team leader for a collaborative writing project? How should the work be divided?

8. List some ways to manage a writing group. What are some good ways to motivate the team?

9. You have developed inventory management software for small retailers. What kind(s) of documentation will you deliver to the buyers of your software? Who would use it?

Chapter 5
Oral Communications

5.0 Introduction

When you contemplate giving a speech, how do you feel? Do you anticipate the sweat pouring down your face and trembling hands, and do you imagine the faces staring blankly back at you while you struggle for words? Almost everyone has some dread of public speaking. But, as an engineer or technical professional, you have lots of information to share – and a place in the spotlight is just what you need to get recognition and advance your career.

Now imagine that you are standing in front of a group of professionals who are interested in what you have to say. You walk to the podium confidently and smile at the audience. You immediately capture their interest by a humorous or intriguing remark. You then explore the topic with poise and well-chosen words, smiling and making eye contact as you look around the room and bring everyone into your circle of speech. At the end of your talk, many hands are raised to ask relevant questions, which you answer assuredly and with good humor.

After such a speech, you are seen as someone who is well-rounded – someone who understands technology AND people. Someone, in fact, who would make good management material or could make client visits with the sales team. Of course, your manager will want to retain you with plum assignments, salary increases, and bonuses!

But how do you turn the dread of speaking into success? Very few people are naturally good public speakers – our instincts are to avoid setting ourselves up for ridicule or criticism. On the other hand, we want to be able to express ourselves to a group of people: to entertain, convince, and educate them. Good public speakers gain their skills through desire, coaching, practice, and hard work.

This chapter covers oral communications, including presentations and meetings. Its sections focus on preparing for a presentation, making the presentation, creating a PowerPoint presentation, and meeting participation. Suggestions are given for increasing self-confidence and keeping the audience interested in what you have to say.

A good speech leads to success

5.1 Principles of Public Speaking

A Dozen Important Principles

We begin this section by focusing on the 12 most important principles of giving effective oral presentations. They are not presented in any particular order since every one of them is extremely important. We list them first and then examine each in detail. Some of these principles are expanded upon later in this chapter. They are:

Principles of public speaking

- Don't apologize

- Connect with the audience

- If you don't rehearse, don't present

- Speak slowly and clearly, and vary your speech pattern

- Avoid bad habits

- Project confidence and enthusiasm

- Don't read from notes

- Project confidence

- Dress appropriately

- Avoid typos on slides

- Listen to others

- Enjoy yourself

1. **Do not make apologies.** The audience will assess your talk based on how well you present it. It doesn't make any sense to influence them with negative comments at the beginning. If you follow the other rules we give, you will not find it necessary to give any excuses. We have all seen some speakers who have a specific inherent problem such as stuttering. Some such speakers start with an apology ("Sorry, I stutter") and then proceed to give a perfect speech without any hint of an impediment. They may believe that the apology has a calming effect, but if they come thoroughly prepared and rehearsed, an apology should not be necessary.

2. **Get to know the audience.** One way to reduce nervousness is to arrive early and talk to a few audience members. Introduce yourself and find out

a bit about each of them and their expectations from your talk. During the talk, you will be able to identify a few friendly faces, and this really helps keep down the tension level.

3. If you don't rehearse, then don't present. This is perhaps the most critical advice we can give. The audience, be it large or small, has given their collective time to come listen to what you have to say. To present without rehearsing is very selfish and shows little respect for the audience.

4. Speak slowly and enunciate. Avoid filler words. Vary the tone, volume, and rate of speed of your speech. Many speakers seem like they are in a big hurry to finish yet the audience is probably hearing this material for the first time and needs some breathing space to absorb it. Pause between sentences and speak as slowly and clearly as you can. Avoid filler words such as "like," "of course," "you know" and "um." You may not even realize you are doing this. Nobody likes a monotone. Try to modulate your voice and add interest to your words.

5. Avoid distracting mannerisms. Don't pace back and forth or move around the room during your talk. Don't put your hands in your pocket, cover parts of your body with them, or hug yourself. You should also avoid looking at the projection screen and turning your back to the audience. If possible, use a computer monitor to look at your presentation. If you turn away from the audience, you will not be able to make eye contact – one of the most important ways to connect with your listeners and to help you gauge their attention. If possible, you should arrange to record one of your presentations so you can see if you have any of these annoying and distracting mannerisms.

6. Project confidence and enthusiasm. If you are not enthusiastic about your presentation, how can you expect the audience to keep interested? You should move naturally, using your hands for emphasis.

7. Don't read from notes. You can have a brief outline or key facts written on index cards, but learn to glance at these rather than looking down and reading word for word. You want to maintain eye contact with the audience throughout the speech

8. Stand erect and breathe deeply – **assure yourself you have something of value to present**. Not only does posture affect your voice, but it also helps reduce your level of nervousness. You have something of value to present to your audience, and you should feel confident. Try not to focus on the negative (e.g., there are some very smart people in the audience who might ask difficult questions).

9. Wear suitable clothes. The culture of the organization will help determine how you should dress. In some cases, it is appropriate to be quite casual. The main rule of thumb is that you don't want your audience to focus on your clothes instead of listening to your talk. Inappropriate outfits can detract from any presentation. Another rule of thumb is that it is better to be over-dressed than under-dressed. You will usually not create any problems if your audience turns out to be dressed in jeans and t-shirts and you show up in your "job interview outfit." The reverse scenario would detract from your speech.

10. Avoid typos or other mistakes on slides. This principle goes along with rehearsing thoroughly. You should spend the necessary time to carefully proofread your slides. Have someone else look though them. Typos or mistakes not only detract from the professionalism of your talk – they also can cause you to lose your train of thought, as you notice them for the first time.

11. Listen to others giving presentations. You can learn a lot from others so take advantage of opportunities to attend presentations. Pay close attention to the techniques these speakers use. This advice also applies to talks within the same program as yours. Sometimes a block of time is set aside in the class or at a company design review, and a series of speakers give their presentations. It may be tempting to stay out of the room until your appointed time, so you can do last minute preparation for your speech. Many audience members will rightfully consider this behavior to be selfish. If others are taking the time to listen to you, you should return the courtesy. Doing otherwise makes it appear that you think you are the main attraction and the other speakers are not important. In addition, by sitting through the other speeches you can often put your speech in context, and you can gauge the response of the audience (e.g., how many questions do they ask, do they laugh at jokes?).

12. Lighten up and try to enjoy yourself. Perhaps even tell a joke. When a presenter is relaxed and having a good time, the audience will enjoy the speech. Livening up the speech with a quip or a story helps keep the audience's interest. Everyone likes to be entertained!

5.2 Giving a Speech

Giving a speech requires preparation, delivery, and perhaps some follow-up.

Preparing the Speech

When you plan to make a speech, there are some choices to be made about its preparation, depending on the mode of delivery:

- You can just stand up and talk about the subject.
- You can write an outline and speak to it.
- You can write a scripted speech to read.
- You can put together slides containing highlights of your speech and also scripted notes for reference during your speech.

Let's explore the pros and cons of each option.

Extemporaneous: If you are very well-versed in the topic, you may choose to simply talk about it without any additional preparation. In Section 5.1, we emphasized the importance of rehearsing. With extemporaneous speeches, rehearsal can be replaced by studying the topic and convincing yourself that you know it thoroughly. Giving an extemporaneous talk saves preparation time, but you will run the risk that you will "clutch" at the podium, forget to say something, or say something you wish you hadn't. In addition, you will have no documentation of the speech, so if there are issues or questions about it later, you won't have any proof of what you did say. It also gives the impression that perhaps you are not prepared, don't know how to manage your time, didn't think the presentation was very important, or aren't professional. The worst thing about this approach is that it erodes your self-confidence since you know you risk failure – and self confidence is critical to good public speaking.

Outline: You can prepare an outline and then reference it to assure that you cover all points in your speech. This approach allows you to choose your words as you go, tailoring your talk to the response of the group. You will not have to do much reading, so your eyes can be on the audience instead of your paper. The outline approach involves some of the same potential problems as the extemporaneous approach. You could forget to say something, or say something you wish you hadn't. You also have very little documentation of the speech.

Script: If you write a complete speech, you can think carefully about each word you want to say – before the pressure of standing in front of an audience. Even though you are reading the speech, you will need to know it well, so you can raise your head and look at the audience while you are speaking. A written speech gives you a record of your talk, so there is no controversy about what you have said. It also gives you confidence that you know what you are going to say. The disadvantages of a scripted speech are that it does not appear at all spontaneous. You will have trouble maintaining eye contact with the audience. If you choose to give this type of speech, make sure there is sufficient light at the podium so that you can read your script easily.

Slides: You can develop a set of slides (see PowerPoint presentations in Section 5.3) to show while you give your talk. You can have notes, an outline, or a script to read as you go through the presentation, so you will not have to worry about forgetting anything. The audience will spend some of the time looking at the screen, so you may not feel as exposed and nervous. You will also be able to check your notes while your audience is looking at the slides, which can consist of phrases, bulleted items, diagrams, charts, spreadsheets, etc. and will document your talk. You can circulate copies of your slides as necessary, either before your talk or later. The major disadvantage of using slides is that it takes more time to prepare the talk. Additionally, if your slides are not well conceived and prepared, they could detract from your presentation.

How do you choose the right approach to developing your speech? If the speech is informal and you are pressed for time, you may decide to use the extemporaneous approach. If, however, it is an important talk and you are trying to impress your audience, you should not use this method. If you have visuals to present, especially charts, graphs, figures, or pictures, you should use them. Although you can distribute copies of your slides beforehand, the listeners are then tempted to follow along on the paper or take notes on it and you can lose their attention.

Successful Delivery

Regardless of the approach you take, successful speakers need to have some tools for delivery. As mentioned above, a key attribute of good speakers is self-confidence, which creates a good impression and raises the speaker's credibility. But what about stage fright? Don't we all feel nervous in spite of our preparation? It's true that even the best-prepared and seasoned speakers can suffer from some level of stage fright. More creative people tend to be more sensitive and vulnerable to stage fright, but they also have more resources to overcome it.

Stage Fright

It might help to realize that stage fright is normal and is a physical response to chemicals in our body. When we are frightened, our adrenal gland produces adrenalin to speed things up in anticipation of the "fight or flight" scenario. It can happen very quickly, and suddenly you can feel your heart pounding, hands sweating, or stomach churning. When you feel it occurring, take very slow deep breaths and try to relax. Deep breathing tells your body that there is no tiger chasing you – at least this time – and it can turn off the adrenalin. Some performers imagine that they can turn (adjust) the valve on their adrenal gland and claim that it really works!

Even if stage fright hits you, try to keep thinking about your speech and preparing yourself for a good delivery. You can be prepared with a welcome, a little quip, or joke. Remember – a little adrenalin is actually a good thing since it gets you keyed up for a good delivery. No matter what you do, it will take awhile for your body to normalize, so try to keep focused on good breathing and relax as you get started on your speech.

If you are in the room when the audience arrives, be sure to greet and chat with members of the audience before you start. It is much easier to talk with friends than to give a speech to strangers. If you feel nervous, don't apologize! Chances are no one in the audience even noticed. Once you get going with your speech and your nervous energy is translated into enthusiasm about your speech, your stage disappears.

Some people use visualization to remove the scary aspects of getting up in front of people (e.g, imagining that the audience is naked, or statues.)

Attitude

When you are scheduled to give a talk, try to have the attitude that you will enjoy the experience and visualize a successful event. If you are enjoying giving your presentation, you will naturally make eye contact, smile, and draw the audience into your discussion ~ after all, don't they want to enjoy this talk too? Be sure that you are well-fed and rested before the speech so that your body is able to handle the stress better. And don't forget to breathe slowly and deeply – your body needs that too. If you feel it is helpful, add some coaching words to your notes: Breathe! Slow down! Smile!

Practice

Practice, practice, practice! Never give a speech without practicing it, ideally in front of people. Of course, with a completely extemporaneous speech, you cannot practice the exact wording. A practice session does the following:

- Allows you to review the words of your speech to see how they sound in front of an audience – and to edit them if necessary
- Provides you with feedback from the audience: Is it too long? What questions do they have? Any suggestions for improvements?
- Gives you an opportunity to try out any equipment or technology you will be using
- Helps you to practice your delivery, building your self-confidence and helping you rely less on your notes

Many people have honed their speaking skills by joining Toastmasters International (see www.toastmasters.org), an organization that provides members with opportunities to hone their skills by delivering speeches as well as listening to the speeches of others. You can also take courses in public speaking, including some that teach techniques such as voice control and poise.

Other tips:

Don't pace. While you can walk around the room, especially to make sure that you connect to all members of the audience, pacing back and forth is distracting and makes you appear to be nervous.

Know your audience:

Don't talk down to them or use jargon and acronyms they may not understand. If you have concerns about what members of your audience want to hear, you can ask them some questions at the beginning of the presentation.

Gestures:

Your gestures should be natural and complement your speech. Don't "talk with your hands" or keep your arms rigid at your side. If you are standing at a podium, don't grip the sides and lean against it since your body will curve and your head will bend over in an unnatural way.

Voice and volume control:

Ask the audience at the beginning of your talk if they are able to hear you and make adjustments to your voice and/or microphone. Of course, you should check your microphone (if there is one) before the presentation if possible. Be sure to breathe and make sure your voice is coming from your diaphragm, not your nose. If you can, record yourself and make adjustments to your voice as necessary.

Props:

Some people find it helpful to have props for illustration or humor. Think about what you can use in your speech to focus your audience on the topic or keep the presentation moving along.

Questions:

Tell your audience if you would like them to ask questions during the talk or if you would like to wait until the end of the presentation. Some presenters like to have questions right away so they can gauge the interest of the audience and use questions to create interactions that reduce tension. If you get a question that you can't answer, just be frank and tell your audience you will have to get back to them later. Then be sure to do just that − get back to them. No one expects you to know everything!

Tips for dealing with questions:

In most cases your speech will be followed by a question-and-answer period. This part of your talk is very important since audience members will often form an impression of your presentation based on your most recent experience. That experience is the question period at the end. So here are our rules followed by some details.

1. Encourage questions.
2. Listen carefully to each question.
3. Repeat each question.
4. Answer to the entire audience.
5. Admit lack of knowledge.
6. Don't spend too much time on a single answer.

1. Encourage questions. Unless the program format expressly forbids a question period, you should encourage your audience to ask questions. Even though this may be a bit scary, it involves the audience and can help clarify any confusing points in your talk. As mentioned earlier, carefully consider whether you want the audience to hold all questions until the end of your talk. It is often best to hold the question period until the end of your presentation. If you permit the audience to interrupt your talk with questions, it can break your train of thought and also make the talk seem disconnected and fragmented. Some will argue that a question may only be relevant right after you have discussed a particular point. However, make sure you know how, when using a PowerPoint presentation, to return to an earlier slide when a question is asked.

2. Listen carefully to each question. After the first few words of a question, you may think you know what the audience member is asking, and you might interrupt the question. Such interruption not only deprives you of a precious few moments to think about your answer, but it also poses the danger of you misinterpreting the question. One exception to this rule occurs when someone asking a question becomes very long-winded. Indeed, some audience members just want to show off their knowledge. With experience, you will learn subtle techniques to cut off such extended "mini- speeches" (and we don't mean looking at your watch and loudly sighing).

3. Repeat each question. This is really appreciated by the audience. There are at least four good reasons to do this. First, it assures that you have heard the question correctly. Second, it assures that everyone in the audience can hear the question. Third, it gives you time to think about the answer to the question. (There have been classic stories of international politicians you use interpreters when being interviewed in English, even though they spoke fluent English. During the period that the interpreter translates the question, the politician has additional time to think about the answer.) Finally, repeating the question involves the entire audience in the question. This last point is so important that we highlight it in the next rule.

4. Answer to the entire audience. Few things are as annoying as having to sit through a private conversation between the speaker and a member of the audience. When you repeat the question, the entire audience can take ownership of that question, and then your answer should be addressed to everyone. Involving everyone keeps them attentive and it also avoids duplicate questions.

5. Admit lack of knowledge. It is quite likely that you will be asked some questions that you cannot answer. Don't panic. If you have prepared properly, you need not be embarrassed when you are unable to answer a question. Experienced speakers have developed techniques to turn such questions to their advantage. They sometimes thank the questioner for asking a good question and promise to find the answer and communicate directly after the meeting. Others ask if anyone in the audience can answer. The one approach that does NOT work is guessing or faking the answer. You could very well get caught, and it will invariably be embarrassing.

6. Don't spend too much time on a single answer. A question can serve as a jumping off point to cover a lot of topics that were not part of the primary speech. While this could be tempting, particularly if you are knowledgeable in this area, it deprives others in the audience from asking questions.

After the question-and-answer period, conclude your speech by thanking your audience for their interest and attention. Mention any follow-ups that you will provide. Be sure to make eye contact as you conclude the meeting and include the entire audience in your remarks. You can mention any other relevant information or coming events of interests.

Ah – a big sigh of relief! Your presentation is over, but remember – the phrase "practice makes perfect" applies to giving speeches and using PowerPoint. You will find that your performance improves with each presentation as you become more comfortable with public speaking.

5.3 PowerPoint Presentations

PowerPoint has become the standard for all engineering and corporate presentations. This Microsoft application can be enhanced with other programs that can create animations, modify photographs (Photoshop), or provide very professional transitions between slides. These extra effects can make a slide show very interesting, and tend to be used for presentations to large, diverse audiences. For purposes of periodic company design reviews, there is usually no need to get overly fancy. In fact, it can detract from the technical content of the presentation. It can also make it seem that you have too much time on your hands!

While not wanting to go overboard, you still want your presentation to look professional. There are some basic rules for creating PowerPoint presentations. We first present the six basic rules and then expand on each one.

PowerPoint rules
- Don't clutter the slide with too much art.
- Don't use complete sentences.
- Use simple phrases and key points.
- Use graphics and colors.
- Match the number of slides to the length of the presentation.
- Avoid typographical errors.

Don't clutter the slide with too much art:

The ability to import clip art or photos from the web into a presentation sometimes encourages people to get carried away. Indeed, it almost makes us look as if we know how to draw! But a little bit goes a long way, so don't fill the slide with lots of cute cartoons and colorful clip art. Use art sparingly to help illustrate a point.

Don't use complete sentences:

One of the biggest mistakes presenters can make is to put too much text onto a slide. Some people even scan and copy the pages of a technical paper. You want the audience to listen to you and not spend their time reading your slides. So just put key points in an outline format. The one exception to this rule is if you have an accent, and people might have trouble understanding you. In that case, it is acceptable to have more words on the slide but not so small that the people in the back cannot read it. An apology such as, "I'm sorry, this slide is hard to read," is not acceptable. Such an apology invites your audience to wonder why you bothered including the slide at all.

Use simple phrases and key points:

Your slide should be an outline to guide the audience, but you want the audience to listen to you instead of read the slide. Therefore, your slides should contain simple phrases that a reader can comprehend at a quick glance. You should avoid too much detail so you don't overload the listener. The people attending your presentation can only absorb a limited amount of information.

Use graphics and colors:

Just as with a written document, line after line of text is boring. PowerPoint gives you the ability to make your slides look professional with use of color and templates. It is usually best to pick from the built-in program features rather than select your own template elements. A lot of experience and expertise has gone into selecting colors that look good together.

Match the number of slides to the length of the presentation:

Few aspects of a presentation are as annoying as the speaker presenting too much material. Some presenters who find themselves in this situation rush through the last group of slides, thereby leaving the audience with a very negative impression. You should have taken the time to figure out exactly what is really important in your presentation, just as you do for your written work. Make sure to rehearse to plan the correct amount of material for the allocated time slot.

Avoid typographical errors:

Computers make spell-checking very simple, though you also need to carefully proofread your slides. Better yet, find someone else to go through your presentation to find any errors. Unfortunately, when errors occur the audience focuses more on these than on the substance of your talk. But if you do find an error during your presentation, don't focus on it. Just quickly mention the required correction or continue without any comments. The more attention you bring to errors, the worse the situation.

PowerPoint advanced features:

Once you have mastered the basic features of PowerPoint, you may want to explore some of the more advanced features. These include animations, slide transitions, and sound.

The most common animations in PowerPoint relate to the way the images appear during your presentation. You can use a variety of interesting effects. For example, if one of your slides consists of five bulleted phrases, you can have each of these fly across the screen into position (each time you click the mouse or press a key). Alternately, you can have each line spiral in as if it were flying, or it can zoom in starting small and getting larger. You can apply these effects individually or to an entire presentation at once. There is an advantage to varying the animations so that the viewer does not get bored with one effect repeating over and over again. We refer you to the help screens (F1) for details on how to do this, since they vary depending on which version of the software you are using.

Another type of animation relates to slide transitions. You can use various effects when you move from one slide to another. The new slide can enter from any of the four sides of the screen and move into position. It can telescope and zoom, fade, spiral, pinwheel, or checkerboard. Experiment with the various effects to see which ones you like. Not every effect is available on all versions of the software. You can also set the program for random transitions, which means the software randomly varies from slide to slide.

Yet another form of animation involves showing a brief video clip as part of your presentation. Here is where things get a bit complex since not all of the formats are supported by PowerPoint. One common approach uses a Shockwave file (.swf extension), which you create using Adobe Macromedia Flash. This file can be easily imported into your presentation using ActiveX control (see help screens in the application for details). If you have a video in a format that is not supported by PowerPoint, you can always pause the presentation and show the video before resuming.

PowerPoint has a library of built-in sounds. Most of them are in the category of "cutesy" such as breaking glass, applause, screeching tires, and a cash register (a younger audience may not recognize the cash register sound – of course they may not even recognize the applause). We urge you to use these effects very sparingly. In the early days of PowerPoint, audiences seemed to enjoy hearing the various sounds. As audiences become more and more sophisticated, you will find that sounds can detract from your presentation.

Logistics:

In many cases, you will not have any control over the room setup, but you should be aware of some basic logistics for your presentation. If you are using a laptop computer, you must be able to see that screen without turning away from the audience -- even though the room may be relatively dark, making eye contact is still important. If you are using notes in addition to what is displayed on the computer, you need to make sure there is enough light at the podium (or desk, or table). You may want to bring a small flashlight as a backup system. It is always helpful to have a remote control for the computer, so you can advance slides without using the keyboard. This gives you mobility, so you are not locked in one position for the entire talk.

Where you stand during the presentation is very important. You must never block the screen. If possible, your audience should be able to see both you and the screen simultaneously (otherwise, your presentation is no more personal than if it were filmed in advance). If you must point at items during the presentation, it is best to do so using the cursor (perhaps change it to an arrow) or a laser pointer. Don't count on being able to point with your hand. The screen may be positioned out of your range of reach. If you use a laser pointer, make sure the beam is wide enough so that people can see it from the back. A very tiny red spot on the screen is almost impossible for the audience to see.

Finally, become familiar with navigating through the slides. If someone asks a question at the end of the talk, you may want to return to one of the earlier slides. If the slide advances prematurely, you want to be able to correct this without getting flustered.

5.4 Organizing Your Presentation

The general structure of a talk should be similar to that of a paper. There are four main parts:

- Introduction

- Body
- Conclusion
- Questions

The purpose of the *introduction* is the same whether you are writing a paper or giving a talk. You want to give the audience some idea of what to expect and why they should care. After hearing the introduction, the listeners should be able to decide whether or not they want to listen to your talk. You want to get their attention and make them want to hear what you have to say. Introductions should not be longer than about 10% of the entire allocated speaking time. If you make it any longer than that, you will probably not have enough time to convey the information in the body of the talk.

The *body* of the talk represents the core of your presentation, where you give the detail about your subject. You should realize a talk is different from a paper. When people read a paper, they have the luxury of dwelling on one paragraph for as long as they wish. They can even take the time to check references to learn related material before continuing to read it. Alternatively, a talk is a highly structured and time-constrained activity, and the listener may not have enough time to absorb everything being said. Unless your talk is recorded, the activity is a form of synchronous communication – both the speaker and the listener need to participate at the same time. Alternatively, communication via a paper is asynchronous, since the reading is not taking place at the same time as the writing.

Because of time constraints, you should avoid lengthy derivations or proofs. In many talks you are really presenting an outline or summary of a larger work, such as a study or an associated paper. In such cases the purpose of the talk is to convey the scope and highlights of the study or paper and help the listeners decide if they want to read the entire work.

The *conclusion* lets the speaker wrap up the talk by summarizing the points that were made. The conclusion should map to the things you talked about in the introduction and should convince the listeners that you have accomplished your goals. The audience should know that your talk is over from the words you choose in the conclusion such as, "In conclusion..." or "To sum up the discussion." A conclusion should not be overly long, since you want the attendees to leave your talk wishing that it had not ended. You should also not make a jarring transition

saying something like, "We will talk more about this later." Maybe the speech drifted off the topic too much or was just plain uninteresting.

We are certain that you don't want to be one of these speakers. You should get applause because the audience is enthusiastic and positive about your presentation. If audience members see your name listed as a speaker at a future meeting, we hope they will remember how good you were and want to return to listen to you again.

How do you assure that people want to hear from you again? Here are some tips for making a good presentation:

- Keep up the pace, matching the length of the talk to the amount of time you can keep the audience interested.
- Be enthusiastic.
- Be humble.
- Have fun.
- Involve the audience.
- Help the audience have fun.
- Have a great conclusion.

Keep up the pace:

No matter how good you are, audience members will start getting restless if your speech moves along too slowly or goes on too long. Most listeners feel better about a speech if they have some idea of how long it is going to be. Often this is set by the structure of the meeting (e.g., 15 minutes for each speech and five minutes for questions and answers). As long as the moderator holds people to the schedule, this assures audience members of predictability of length. If the time is more open-ended, you should try to give the audience an idea of length at the beginning. One experienced speaker carries this to extremes. At a recent speech, he started by saying, "My speech tonight is 748 words and will take about seven minutes and 32 seconds." While the speaker got laughter from the audience, it is certainly not necessary for you to be so precise.

You might want to say something like, "For the next 15 minutes, I will be talking to you about ....." You can tell if you are speaking too slowly or for too long if you see audience members looking at their watches. We hope you never experience audience members getting up and walking out of the room, but remember that there could be a good reason other than your speech. If you are speaking at a conference with many simultaneous presentations, people commonly leave after getting the gist of a talk, so they can attend another one. Don't be distracted by this and don't make any reference to it in your talk. "The show must go on!"

Be enthusiastic:

We can't emphasize this enough. If you are not excited about your topic and the opportunity to talk about it, how can you expect the audience to be? Some companies that present short courses and seminars pay their instructors based on the class evaluation. This may not seem fair when the presenter had nothing to do with writing the course or preparing the PowerPoint presentation. But, in fact, the presenter can make all of the difference in influencing the audience. A skilled presenter can turn even a poorly written course into an interesting and instructive one. Have you ever had a professor who criticizes the assigned course textbook? The impressionable students invariably develop a negative feeling about that book (plus the professor who more than likely made it the assigned text.)

Be humble:

Elsewhere in this text we suggest that before you give a speech, you convince yourself you have something important to say. But this can be carried too far. You must avoid giving the audience the impression you think you are the profession's greatest gift to the public. Nobody likes someone who is pompous, and part of your goal should be to get the listeners to like you. If they like you, they are probably going to listen to you. Just because you are giving the speech doesn't mean you are smarter or more capable than the audience – you have simply spent focused time looking at a particular subject.

Have fun giving the speech:

This tip is related to displaying enthusiasm. If you appear to dislike giving the presentation, or your body language (hands in pocket or nervous actions) and expressions (sighing or apologizing) indicate you wish you were not there, the audience is sure to react negatively. If you convince yourself you are going to enjoy giving the presentation, you will automatically display more enthusiasm and confidence.

Involve the audience:

The big push in education is toward *active* rather than *passive* learning. When the learner participates in the process, the result is always more positive. When listeners are involved in your talk, they will better absorb what you are saying, and the time will fly by for them – so you will be very successful. Involve the audience any way you can. One way is in the manner in which you answer questions. Make sure to answer to the entire audience. Some speakers feel comfortable asking the audience a question during the speech. If you do so, it is best to ask something that can be answered with a show of hands (e.g., "How many of you have ...."). You can also invite the audience to ask questions as you go along. (If you do this, be sure to be aware of the time you are spending on such questions)

You should also use eye contact during the talk to make all audience members feel involved. Also, do what you can to relate the topic to the interests of the audience. If you have taken the time to get to meet some of the people before the start of the speech, don't hesitate to involve them. You can say something like, "Mary (point to her) told me before we started that your laboratory is very interested in learning what others are doing in the area of .... I hope my talk will help clarify some of those issues."

Help the audience have fun:

It's not enough for you to have fun. You should try to make your talk fun for the audience. In the extreme, you can start off with a joke if you feel comfortable doing so. Make sure that your joke is relevant and not offensive, or this approach will backfire. Body language and facial expressions can really help draw the audience into your talk and get them involved. It's amazing how infectious a smile can be (try it as you walk down the street). If something goes wrong with your speech, use the opportunity to make a joke or remark about it. Your audience wants you to succeed, so they will just laugh with you.

Have a great conclusion:

The conclusion of a speech (or a paper, for that matter) wraps it up and is very important! The first thing it should do is make it clear that you are finished. Porky Pig had one way of letting you know the end had come ("That's all folks"). Some speakers have adapted this and end their talks by saying something abrupt like, "Well, that's the end of my talk." Please don't do that! It is a terrible way to end a talk and is only marginally better than stopping in the middle of a sentence and walking off the stage.

The introduction and conclusion of your talk are its bookends. The conclusion gives you a chance to point out to the audience that you have accomplished what you set out to do in the introduction. You could say something like, "At the beginning of this talk I told you that I intended to bring you up to date on developments in the field of …. We have talked about x, y, and z, which are all emerging technologies with tremendous implications for our field. I urge you to watch the literature for further developments." You can also hand out copies of your talk, slides, or paper during the conclusion.

Many speakers conclude their talks by saying something like, "And now I'll do my best to answer any questions you may have." This is better than simply saying something like, "that's the end of my presentation," but it is certainly not as effective as it could be. Your conclusion should reinforce the impression that you have effectively communicated everything you promised. You can also invite audience members to communicate with you in the future if they have questions or additional comments.

Whatever you do, don't give a long conclusion. Once it is clear that you are wrapping up your talk, your listeners will be expecting it to end. An overly-long conclusion makes people restless – keep it short and sweet!

And that concludes this section (sorry! We just wanted to illustrate what *not* to do).

5.6 Meetings

As a technical professional, you will probably be collaborating with many people and groups. An effective way of sharing information, making decisions, and gaining agreement is in a meeting. You will attend many different meetings, as both a leader and as an attendee. Here is a list of some typical meetings for a technical professional:

- Status meetings
- Project review meetings
- Work group meetings – accomplish a task (divide an organization, document process flow, perfect a technical approach)
- Interviews with user groups
- Training sessions – as a student and/or teacher
- Client meetings
- Negotiations
- Performance appraisals

Most of your meetings will be within your organization. External meetings can require more formality and attention to detail (handouts, refreshments, etc.)

Successful Meetings:

What constitutes a successful meeting? What is the difference between good meetings and bad meetings? What makes a meeting interesting? What makes people want to come to meetings and happy that they did? How do you motivate people to attend a meeting?

A successful meeting is one that effectively and efficiently accomplishes its goal by a collaborative effort among members of the group. Such a meeting is well-planned and executed, sometimes with the help of a facilitator, who is there to guide the discussion. Participants who are prepared for the meeting and are able to contribute are satisfied that their effort was recognized and are happy that they attended. In fact, many people think a meeting was a success if they have had a chance to talk and be heard. Most people are motivated to attend and participate in a meeting if they care about the issue and want to influence the outcome or be part of a solution. Everyone wants to be valued and appreciated.

Tips for Meeting Leaders:

- If you are calling a meeting, make sure that it is really required; no one likes to waste time in an unnecessary meeting.
- Issue announcements of the meeting place and time in advance and state the goal of the meeting; express your appreciation of invitees' attendance and participation.
- Take others' schedules into account before choosing a date and time.
- Make sure the meeting is documented. If you can't do it, designate someone else ahead of time to take accurate notes and issue minutes.
- Schedule an appropriate room and make sure that necessary supplies are available, such as projection equipment, paper, flip board, markers, etc. You certainly don't want to waste everyone's time by looking for these things at the start of the meeting.
- State everyone's name and role in the organization or project either by formal introductions or by having each participant introduce him or herself. If the meeting is large, you can introduce the participating groups along with the reason they are invited. Pass around a sign-in sheet so you will know later who attended the meeting.
- Start the meeting on time and keep it on schedule. When you keep the meeting moving, it is more interesting.
- Review the meeting goal at the beginning and end of the meeting so the group feels successful when the goal has been accomplished.
- Develop an agenda based on the goal by breaking the task into steps or a discussion into parts.
- Let the participants shape the agenda, and make sure they all have a chance to participate. Using humor can help everyone relax and feel comfortable speaking up in a group setting.
- Use your agenda to keep the meeting on track. If a side issue comes up, ask participants if it needs to be discussed. If the discussion veers away from the main topic or gets into

When an external group is involved in a meeting, you need to understand the goals of the meeting and your role in it. A project status meeting with the client for your software development work is not the time to bring up schedule problems with the project manager. You are part of the corporate team presenting a united front to the client.

Meeting Follow-up:

It is a good idea to email a meeting summary to all participants, especially if there are resulting action items. A meeting summary can:

- Summarize topics discussed
- Document suggestions and decisions
- List action items and those responsible
- Provide follow-ups and due dates
- Thank everyone for attendance

A successful meeting gets everyone involved

5.7 Interviews

An interview is a dialogue between two people (sometimes more), with the interviewer asking questions of the interviewee. The interviewer asks questions to obtain information in order to make an assessment, as in a job interview. First we will look at the role of the interviewee in several types of interviews; then we will discuss conducting an interview as a professional.

Job Interviews:

A common interview is the job interview: a meeting between an applicant (or "candidate") and a prospective employer to determine if there is a good fit between the two. Usually an interview is scheduled after staff in the hiring organization reviews the applicant's resume and cover letter. For a specific job opening, the normal flow of the application starts with the personnel or human resources group, who check the resume against the job requirements. The resume is then normally sent to the person or team for whom the candidate will work to decide whether or not to invite the applicant for an interview.

Campus recruiting takes place through the school; organizations schedule interviews on campus, often in the fall, to determine who will be invited to the company's site for extensive interviewing. College candidates are not usually targeted for a specific division in an organization, but are hired to fill entry-level vacancies wherever they are needed when the candidates come on board.

Job interviews can be stressful, but they can also be informative and exhilarating. After all, candidates have made it through the resume assessment process, and it is really nice to be wanted! Successful interviewers look at the interview much as they would a lunch date: an opportunity to get to know the other person and a dialogue rather than an oral exam or a one-way question-and-answer session.

Campus Job Interview:

When companies come to schools to recruit newly-graduated technical professionals, an interviewer will typically talk with eight to ten people in a day for about half an hour each. The interviewer will document each interview and decide which applicants should be invited to more extensive interviews at the corporate location. Usually a representative from the company will meet with prospective employees before the scheduled interviews to give an overview with general information, so interview time will not be wasted. Attending the meeting is crucial to having a successful interview, since you will learn about:

- Timely information about the company and its background

In addition to the information you are giving the interviewer, your answers show if you can think on your feet, if you have good interpersonal skills, if you make good choices, and if you are mature.

On-Site Job Interview (second round of interviews after campus interview):

The same rules apply for the on-site interview, but you will usually have more time to make an impression and you will be speaking to more than one person.

Job interview advice:
- Arrive on time. When you are interviewing in an organization's office, you will be in unfamiliar territory, so it is even more important to allow plenty of time to get to the office. If you add the stress of getting lost or running late to the pressure of an interview, you will not be relaxed and at your best.
- Be professional. Dress in your best professional clothes, even if the group you will be interviewing has a "business casual" workplace. You will look and speak more professionally and will show your respect for the opportunity.
- Relax (think about your breathing if it helps) and smile. You need to put the interviewer at ease too!
- Be interested; ask good questions about the company and the position. Take a few notes during each interview, which shows your interest and will help you write follow-up letters.
- Furnish information that the interviewer requests, but make the interview a dialogue. After all, this isn't just about the company wanting you; you may need to make a big decision that will affect the course of your career, so getting information is important to you too. Be aware of questions that are generally not allowed, such as those concerning age, health, and personal information. If you are asked something that is not allowed, just ask why the interviewer wants that information or tell them that they shouldn't be asking it. Don't answer questions about your salary requirements. These questions should be saved until you are talking with the HR department or when you have a job offer.

Lunch Job Interview:

Candidates are usually taken out to lunch when they spend a day in interviews. Continue to speak and behave in a professional manner; you are still being interviewed by everyone at the lunch meeting. Use your best manners: wait for everyone to be served before you begin eating. Don't put your hands on the table, don't look down and stuff in the food, take small bites and chew carefully, and swallow before talking. Remember to smile and use eye contact. If you ensure that the other person is having a good time, you will too!

Phone Job Interview:

Sometimes, a phone interview is conducted for additional screening before the organization schedules office interviews. When you are participating in a phone interview, resist the urge to lean over your notes and answer questions. Keep your head up so your voice will be just as clear as it is in an in-person interview.

Job Interview Follow-up:

After you have had a job interview, it is always appropriate to send an email or written thank-you note. The note should:
- Thank the interviewer for the opportunity to discuss the position and learn more about the company
- Mention something that you learned in the interview
- Give a reason or reasons why you are interested in the position or are a good fit
- Express the desire for follow-up or the next step in the process

A thank-you note makes the interviewer feel more positive about the interview and more inclined to give you a good review.

On-the-Job Interview:

When you are in a professional position, you may receive a request for an informational interview. For example, if you are in charge of a team of engineers working on a design project, a systems analyst may wish to talk to you about your group's requirements for a new software application. If you are interviewed for information, be sure that you understand why you will be interviewed and what questions will be asked. Many interviewers will send you the questions ahead of time, allowing you the opportunity to gather materials to support the interview. Be sure to take notes so that you have documented what you told the interviewer. Sometimes interviewers are so busy interviewing that they make mistakes transcribing your answers. If you are concerned about misrepresentation, ask the interviewer if she or he can send you a copy of the draft documentation.

Conducting Informational Interviews:

If you are conducting informational (or information-gathering) interviews for your job, you will normally be provided with a list of people to interview. For example, if you are developing functional requirements for a new data base system, you will interview representative users, managers, and IT staff.

Preparing:

Developing a good set of questions is critical to a successful interview. Determine the information you need and formulate questions that are detailed enough to get the information, but also some open-ended questions so the interviewer can tell you about things you might have overlooked. Think about sending a copy of your questions or information that you need so your interviewee can prepare for the meeting.

For example, if the interviewer has a list of reports that are typically produced, ask to get a copy. Be sure to note if there is additional follow-up to the interview. If you interview more than one type of professional, you may need to develop other sets of questions.

Introduce yourself by giving your name and job title or position in the organization. Be sure to smile and put the interviewee at ease. Explain the purpose of the interview and provide information about the position. You can review the questions you plan to cover or can carry on a more casual conversation by asking questions depending upon the interviewee's answers. Also ask the interviewee if you can contact him or her later if additional questions arise, and find out whether you should use the telephone or email. Remember that you are taking valuable time to conduct an interview, so try to be efficient.

Scheduling:

You must schedule interviews ahead of time and allow sufficient time for a thorough interview. Don't schedule interviews too close together; you will need time to document each one while it is fresh in your mind.

Documenting:

You should document your interview right away, while it is still fresh in your mind. If you were given materials by the interviewee, review them right away too, so you can get back to the person if something is missing or if you have additional questions or requests.

Follow-up:

It is always courteous to send an email expressing your gratitude for the interviewee's time. A note takes just a few minutes, but shows that you appreciated the interview and that it was helpful. If something arises later and you should need to contact the person again, he or she will be more disposed to help you.

5.8 Leaving Voicemail Messages

You might wonder why we include a section on voicemail messages. In fact, many people don't know how to leave effective phone messages. Your voicemail messages can affect what people think of you, whether you are a student or a professional engineer. You will also find that many of the principles of effective voicemail messages are the same as those of other forms of communication, both oral and written.

How many of you have received a very long voicemail, causing you to sit through many minutes of the message only to find that in the final moments, the caller rapidly leaves a phone number for you to call – so rapidly and after you are starting to drift off to sleep—you miss it? Unless your system allows you to fast forward, you must decide whether to replay the entire message just to catch the phone number or simply to ignore the entire message.

skills are equally important for studies in either field? How do you think your advanced studies would benefit by improved oral communications skills?

5. Prepare a PowerPoint presentation summarizing the key points of this chapter. Then carefully critique this presentation and list ways you could improve it.

6. Regarding presentation skills:
    a. What are some good strategies to use in preparing for a presentation?
    b. What are some good ways to deal with stage fright?
    c. What are other considerations to ensure that you give a good presentation?
    d. List the important things to remember about your voice when giving a speech.
    e. List the major components of your speech.
    f. List the things you like and dislike most about making a presentation.

7. Find some good web sites that provide tips for presenters. Which are your favorites? Why?

8. List methods to interject humor into your presentation. List some websites that are good sources for humor.

9. Prepare a ten-slide PowerPoint presentation describing your favorite hobby. Do this first without using any color or animation. Then modify your presentation to add artwork, color, and animation.

10. Prepare a PowerPoint presentation describing the last vacation or trip that you took. If you don't have any photographs, try to get some from the Internet.

11. Use a search engine to locate three PowerPoint presentations on the web. Choose any topics of interest to you (hint: You can

include the words "PowerPoint presentation" in your search term, and then look for URL addresses that end in "ppt"). Download these presentations and critique them. Discuss ways you think they could be improved.

12. Go to the website for the university you are attending (or previously attended). Search for class notes from professors in your discipline. Try to locate PowerPoint presentations faculty have placed on their websites, and view several of them. What did you like about these? What do you think could have been improved?

13. Assume you are interviewing for a career position and the interviewers have asked you to prepare a PowerPoint presentation summarizing your education and accomplishments. Prepare that presentation. Then identify at least five ways it could be improved.

14. You are a project leader for a new software release scheduled in two months. You need to know the status of each of your five development groups. Would you call a meeting? Why or why not? If you have decided to call a meeting, whom will you invite? How? What will you tell them? What, if anything, will you ask them to bring? If you have decided not to call a meeting, how will you get the information you need? What will you ask your groups to do?

15. You are responsible for the installation and maintenance of your company's network. You have been invited by your manager to attend a vendor's presentation for a new network hub. Will you attend? What, if anything, will you do to prepare for the presentation? What will you do before, during, and after the presentation? Should you bring anything with you?

16. You work for a network hardware vendor as technical support for marketing. You have been invited to make a presentation of your new hardware products to a potential customer. What will you do

to prepare for this assignment? What will you do before, during, and after the presentation? What should you bring with you?

17. You have been invited to attend a software design review meeting by the head of the software development group. You are in charge of the software test group. What will you do to prepare for the meeting? What kinds of information do you expect to receive in the meeting? What kind of information will you need to provide? What kinds of questions will you ask during the meeting?

18. Name five people who you think are good presenters and list the reasons why you enjoy their presentations. You can choose teachers, actors, television personalities, or any professional.

19. Suppose that you are a mechanical engineer in the automobile industry. You have spent the past five years working on designs for production vehicles, but would like to be assigned to work on an exciting new prototype. How would you make this wish known to your manager? How would you prepare? Write what you would say in conveying your message.

20. You are a systems analyst working on a project to automate a paper-based process. You will be interviewing the user community to determine functional and systems requirements. What information will you try to obtain in the interview? Design an interview questionnaire for one of the persons you plan to interview.

21. Attend a technical speech within either your company or university. Write a 200-word critique of the presentation along with suggestions for improvement.

22. Create a PowerPoint presentation "selling" any theorem, chemical formulation, or law of physics of your choice. Your audience is a group of majors in the humanities. Make sure that your presentation will last for at least five minutes without questions and answers. Try to make the presentation interesting and convincing.

23. Voice Mail
   a. Compose a voice mail greeting for your office telephone.
   b. Compose a voice mail message for your hardware vendor to check on the status of a shipment of network routers.
   c. Compose a voice mail message for your manager informing her that there is a problem with the shipment of the routers.
   d. Compose a voice mail message requesting technical support for your PC, which has just crashed – you have a BIG proposal due tomorrow!

24. Career Plans Speech

   a. Write the outline for a short speech about your career plans (five-ten minutes in length).
   b. Write the speech by filling in the outline.
   c. Put together a set of slides or handouts for the speech.
   d. Present the speech to the class.
   e. Solicit suggestions from the class or your peers for improving the speech.
   f. Incorporate comments as appropriate and give the speech again.

25. Give a short talk about a diagram, chart, or other graphic from a technical course or reference. Solicit questions from your audience.

26. Read a professional text to the rest of the class. Be sure to make eye contact and pay attention to your breathing and voice, both quality and volume.

27. Classmates should pair up for this assignment:

   ▪ Develop a PowerPoint presentation on a professional or technical topic that is familiar to you. Review your presentation with a classmate and incorporate comments. Now give the presentation to your classmate and get comments on your delivery style.
   ▪ Present it to the class and solicit comments.
   ▪ What did you learn from this exercise?

28. Your class is divided into two groups, and each group must determine a meeting leader. Each group will conduct a meeting for half of the class, while the other group looks on silently. The meeting will consist of a discussion of engineering and technical topics chosen by the group such as technical career choices, engineering vs. management, graduate study options, etc.

29. Consider the alternate approaches to presenting a speech. Which are the best for your technical field? Why? Do you see any advantages or disadvantages in distributing copies of your slides before your talk? How about after your talk? How about emailing copies of your slides?

Chapter 6
Rules and Tools

6.0 Introduction

This chapter briefly reviews some basic guidelines for good writing and speaking. It is not a comprehensive handbook on English, nor is it a style guide. The appropriate books or web sites can be used for reference when you encounter a question or wish to improve your communications skills further. This section, however, discusses those "rules and tools" that you will use most frequently, especially those that will help you differentiate your work when in a professional position. It discusses:

- Rules of grammar
- Style and format
- Word processing software
- Other writing tools

6.1 Grammar

Oh no, the dreaded topic of English grammar! Engineers sometimes abhor the rules of English grammar, probably because they are viewed as illogical. Grammar is not like math or science. If, however, a few rules (and exceptions) are noted and used in everyday speech and writing, they can become automatic. We will not provide extensive English grammar rules in this book, but will review some of the more common ones and also point out the areas that present the most common problems.

A grammar is a set of rules for constructing language. Programmers use grammars when they write computer software programs. English, however, is different since it is constantly evolving. Spoken language is regional and fleeting, so poor grammar and word choices can be easily overlooked until they almost sound acceptable. For example, a writer could put his or her everyday language into words and think that it sounds just fine (for example, "Me and Jim went to the convention.")

Written language, especially business language, is more static; after all, it is expected to apply to a larger group of readers and it will last for a longer time than the spoken word. Remember that the goal of good writing is to be easily understood and so should be easy to read. Mistakes in grammar and syntax are jarring to the reader and can cause the eye to stop and notice them. Also, such mistakes can create ambiguities and misunderstandings. Consider the sentence below:

Engineers sometimes abhor the rules of English grammar, probably because they are viewed as illogical.

The writer, and most of the readers, may have in mind common knowledge that English grammar rules are viewed as illogical, but take another look at the sentence. What does "they" refer to – rules of grammar or engineers? The sentence would be better if it were rewritten:

Engineers sometimes abhor English grammar, probably because they view the rules as illogical.

Now there is absolutely no confusion about the meaning of the sentence.

There are plenty of style guides for writers (see references), but it is difficult to stop the flow of writing to consult them, so try to read them ahead of time. A perennial favorite is Strunk and White's *The Elements of Style*. Some entertaining ones are *The Elements of Style (illustrated)* and *Eats Shoots and Leaves*. There are also many good web-based resources for checking your grammar usage; some even include exercises and self-tests.

The important thing to remember is that writing should be clear, straightforward, and accurate. The writer should tell a story so that the thoughts progress logically without confusing the reader and without unneeded words. In fact, good writing makes absolutely sure that the reader will not misunderstand – remember this rule when making choices about wording. For example, if you are talking about a specific item, do not use a synonym that would confuse the reader – just use the item's proper name.

Rules of grammar do have some underlying logic, and complying with them makes your writing easier to read and understand. Following are some of the more common definitions and rules.

Parts of Speech

There are eight parts of speech (or types of words as well as word phrases):

- Noun: Person, place or thing −the subject or object of a sentence
- Pronoun: Takes the place of a noun: he, she, they, it
- Adjective: Modifies a noun
- Verb: Describes action or equivalence
- Adverb: Modifies a verb, adjectives, and entire sentences
- Conjunction: Joins words, phrases, or sentences: but, so, or, because, however
- Preposition: Comes before a nouns and joins it to another part of the sentence: in, on, by, with, through, at, under
- Interjection: Expresses emotion or surprise and often stands alone, usually followed by an explanation mark: Hello!, Hey!, Ouch!

Word Agreement

Good writing and speaking requires choosing the correct form of a word, whether to agree with another word or to accurately reflect its role in a sentence. Errors in subject/verb agreement, for example, can cause confusion. Following are some guidelines for determining the correct forms of words to use in various cases.

Pronoun form: A pronoun must take the correct form depending upon whether it is a subject or object. Do not be misled by compound phrases; just drop everything but the pronoun to check for the correct form:

> *The team and I celebrated a successful acceptance test.*
> (You would not say, "Me celebrated a successful acceptance test."
> *The customer presented the award to Jim and me.*
> (You would not say, "The customer presented the award to I.")

Pronoun-noun agreement: Pronouns must agree with the noun to which they are referring: a singular pronoun for a singular noun, plural pronoun for a plural noun:

Plural and Possessive Nouns

Mistakes in forming plural and possessive forms of nouns may seem unimportant, but they pop right out of the page to a reader – especially those that involve an apostrophe. They may not be found by a software tool, so it is best to learn the rules by understanding the logic and/or memorizing them. (Do NOT add an "'s" to a noun to form a plural – the apostrophe indicates possession.)

The plural form of most nouns is formed by adding an "s":

members, bridges, tools

For nouns ending in *s, x, z, ch*, or *sh*, add *es* (which makes it easier to pronounce):

bosses, boxes, peaches

If the noun ends in o, add an *s* or an *es*, depending on the word. There is no rule for deciding this, so you just have to memorize the forms – or look them up if you are unsure:

s plurals: *shoes, memos, radios*; *es* plurals: *tomatoes, potatoes*

If the noun ends in y, drop it and add *ies* to form the plural:

companies, batteries

There are some nouns that take an irregular plural form: man – men, woman – women, person – people, matrix – matrices. Some nouns based on Latin words use the Latin plural form: datum – data, fungus – fungi, medium, media.

Recently, there has been a trend that presents a possible exception to the rule of not adding an apostrophe to form the plural when you are using numbers, letters, or words signifying spoken words:

Seven 10's were shown on the graph. The aye's have it.

Possessive Forms

The possessive form of a noun conveys that it *has* something:

The computer's memory was upgraded. The companies' goal was to win the project. The engineer's mantra was "fast, good, or cheap – you can have any two!"

Now you can use those apostrophes you have been saving! When a noun *possesses* an apostrophe, it is *possessive*. The apostrophe is placed differently for singular and plural nouns.

Form the possessive form of a singular noun by adding an apostrophe followed by an *s*:

the engineer's career, a bridge's capacity

If a noun ends with *s, x* or *z*, omit the *s*: the mass' speed,

Form the possessive form of a plural noun ending in *s* by adding only an apostrophe:

networks' cost, tools' common features

Form the possessive form of a plural noun not ending in *s* by adding an apostrophe and an *s*:

people's goal, data's accuracy

For compound nouns, where there is joint ownership, form the plural by making the last noun plural:

Marty and Terry's goal, the data center and network's cost

If each noun in a compound noun has individual ownership, form the plural for each term:

Marty's and Terry's degrees

Remember that a possessive pronoun (his, her(s), their(s), mine) needs no apostrophe:

The team's laboratory was all theirs.

Articles

An article is used before a noun (*a, an, the*). Use **a** to indicate one of many, **the** to indicate a particular item:

project manager has signed off on the deliverables, so we can now submit an invoice.

Rather than:

The project was successful due to the actions that were taken. The next steps can now be completed.

Sentences: Make sure your sentences don't *run on* (have too many subjects and verbs); such sentences are confusing to the reader. Make sure they are complete, containing both subject and verb.

Using the above example, look at this run-on sentence:

Our team has completed the project which was successful due to meeting its deadlines and was also within budget, so that the user project manager has given us a sign-off.

A sentence with parallel structure means that it contains two or more parts with the same level of importance (words, phrases, or clauses). Parallel structures are usually joined with coordinating conjunctions such as *and* or *or*:

Computers are used extensively in medicine, and they have radically altered every aspect of patient care.

Active vs. passive voice: In active voice, the subject of the sentence performs the action specified in the verb:

Mechanical engineers designed the parts and assembly process for the new engine.

In passive voice, the subject of the sentence is acted upon:

The parts and assembly process for the new engine were designed by the mechanical engineers.

As you can see, the second sentence would still be a sentence if the phrase "by the mechanical engineers" were absent, but would be lacking in a specific detail. Sentences written in the passive voice are often lacking some details and invite further questions. Passive voice obscures the actor; do not use it unless you mean to do so. Sentences in the active voice are always stronger and easier to read.

For example: *Computer Concepts Corporation develops leading-edge software for the telecommunications industry using IEEE software engineering guidelines.*

Quotations: Set off quotations with a comma and quotation marks. A quotation within a quotation is set off by a single quotation mark:

> *The manager said, "Remember the old adage: 'let sleeping dogs lie' and you will avoid problems."*

Long quotations should be indented and single-spaced, eliminating the need for quotation marks.

Italics: Italicize foreign words and titles of publications:

> Please refer to *IEEE Recommended Practice for Software Design Descriptions* for details about our standard documentation format.

Numbers: Spell out numbers under 11. If you start a sentence with a number, write it out even if it is greater than 11.

Set off persons being addressed with commas:

> *Please, Mr. Bailey, consider using a larger disk drive on your server.*

Paragraphs: A paragraph is a grouping of related sentences. Paragraphs usually open with a topic sentence, which is explained or expanded upon in the paragraph. They can also develop a theme or advance a discussion. Often an overly long paragraph, or run-on paragraph, occurs when the writer attempts to develop more than one theme. The paragraph can be split up or shortened by breaking out the different themes. Remember, a paragraph does not have to be long; it can have just a sentence or two, although paragraphs consisting of a single sentence are generally avoided (except in letters and memos).

When the objective of a paragraph is to prove a conclusion, it should usually begin with a statement of the conclusion as the topic sentence. It should then be followed with the reasons why the topic sentence is true: C, A → B and B →C, therefore C. Because of their mathematical training, many engineers start right out with their logic statements: A → B and B →C, therefore C. Readers can be confused, become annoyed, or lose interest when they are confronted with a list of logical statements. After all, they don't know why they should be interested and can also lose the flow of the reasoning before the statement of the topic sentence. It is better to state the conclusion right up front, so you capture the reader's interest and put your arguments in the context of the conclusion.

For example:

The structure of the two databases is incompatible, so we can not easily obtain a merged report of the corporation's data. Because we don't have a view of corporate data, it is impossible to make timely financial decisions.

Consider the alternative:

We recommend converting both databases to a common format. The current incompatibility of the two data formats makes it impossible to obtain good overall reports and make good financial decisions. Once the databases are converted, corporate reports can be obtained in a matter of minutes.

Notice that the second example above begins with the recommendation, followed by the reasons for the recommendation. The paragraph ends with a restatement of the recommendation with its consequences. Isn't it easier to understand the author's objectives?

Sentence Structure

If a paragraph is composed of only simple sentences (subject, verb, object) it will be monotonous and unpleasant for the reader. Vary your sentence structure and link sentences together with appropriate conjunctions.

Look at the following paragraph:

Engineers are needed for our quality assurance team. Quality assurance people test the software. QA people make sure that the software is delivered on time. Good quality is important. Customers need to be able to rely on the software.

It consists of simple sentences without linkages. Now look at the following paragraph and compare the flow:

The Quality Assurance team needs engineers to test the software. Because of the QA team, software is delivered on time with the quality that is important to our customers – who rely on our software.

Lists

If you are including a list in your writing, be sure the numbering is consistent with other lists. Bulleted lists are a good method for focusing the reader on many items and add variety to your writing. Also be sure to use comparative structures for the list items: all items singular, all plural, all sentences, etc.

Confusing Words

Some words can be easily confused because they are so similar.

Affect vs. effect

To *affect* is a verb meaning *have an influence on:*
    The RFP affected the technical approach in our proposal.

Effect can be a noun or a verb. The noun means the resulting outcome:

    The effect of the customer's design change was a delayed delivery date.

When you affect something, you have an effect on it. Effect can also be used as a verb, meaning *to create*:

>*We are trying to effect a change in installation procedures.*

Effect is also used in the phrase "personal effects."

Compose vs. comprise

Comprise means *is made up of* or *consists of.* The whole comprises the parts.
Compose means *made up* or *made.* The parts compose the whole.

>*The compound comprises four chemicals.*
>*The compound is composed of four chemicals.*
>*Four chemicals compose the compound.*

It is incorrect to say: *The compound is comprised of four chemicals.* In general, do not use *of* with comprise since it is already inherent in the term. If you are confused, you can always say,

>*The compound is made up of four chemicals* or *Four chemicals make up the compound.*

Many vs. much

Many is used with a noun that refers to things we can count.

>*Many of the system users were not well trained.*

Much is used for things we do not count.

>*Much of the trouble was from unskilled workers.*

Less vs. few

Few is used in the same way as *many*, with a noun that refers to things we can count.

>*Few of the system users really knew what they were doing.*

Less is used just as *much*, with nouns we do not count.

>*There would be far less trouble if we trained the users ahead of time.*

Lie vs. lay

The verb *to lie* is an intransitive verb: it describes the action of a subject but has no direct object. It is action but not the kind of action that is done to something else. It can be thought of as equivalent to the verb *to recline* or *to rest*. Its past tense is *lay* and its past participle is *lain*.

For example:
> *The patient lies on the table.*
> *The patient lay there yesterday.*
> *He will lie there tomorrow.*
> *He is lying there now.*
> *I have lain there every day.*

The word *laid* is never used to describe the act of reclining.

The verb *to lay* is a transitive verb: it describes the action of a subject on an object. Something in the sentence must be receiving the action. *To lay* means *to put* or *to place*. Both its past tense and past participle are *laid*.

For example:
> *I lay my briefcase on the desk every morning.*
> *I laid my briefcase on the desk this morning.*
> *I will lay my briefcase on the desk tomorrow.*
> *I am laying my briefcase on the desk now.*
> *I have laid my briefcase on the desk every morning this year.*

Because the past tense of *to lie* is the same as the present tense of *to lay,* it is easy to be confused between the two verbs. Another problem is that we are so used to hearing the incorrect form of these verbs that we no longer notice it. Just be aware that the two verbs are often misused, and make sure you pick the correct one.

Tools

Most word processing systems have spell check or an indication that a word is misspelled as it is typed. Always use spell check, but understand that only you can identify words that are not misspelled, but used incorrectly. Grammar software is available as are Internet-based grammar tools, but it is always best to know the rules yourself. Remember that you will be using good grammar not just in your written reports, but also in your emails, conversations, and presentations.

6.3 Word Processor Fundamentals

Word processors can vary, but there are common features among all of them that will make your writing life easier. In the following sections, we present the basics of word processor applications. This should not be considered a manual, because the specific operations vary from word processor to word processor -- and even among the various releases of a particular piece of software. Instead, we just want you to be aware of their major functions and how they can help you in your writing.

Fonts and Font Formats

You can vary fonts and typeface to make sure that certain parts of your document stand out. The most common font used by technical writers is Times New Roman. But sometimes writers use Arial or even Courier. A major difference between the basic fonts is whether they are *serif* or *sans serif.* The serif is the little "blob" at the end of strokes of a letter. For example, if you look at this sentence, which is in the serif font Times New Roman, you can see little hooks at the end of letter strokes. **On the other hand, if you look at it in Arial (the sans serif font we have used for this sentence), the hooks are gone, and we refer to this as sans serif (in French, "sans" means without).** The sans serif fonts are often used for titles, while serif is more common in the body of a paper. On the Internet, sans serif dominates since it appears cleaner on a monitor screen.

Each font has a variety of options. You can select any portion of a document and change it to italics, bold print, or underlined text, combining these in any permutation you wish. Another useful feature used in technical documents is subscript and superscript. For example, in x^2, we have used italics, and the *2* is set as superscript. You can use italics, bold, and underline to emphasize portions of your text. These features are used in the conventions for titles and references. You can also choose different colors for portions of your text. This should be done carefully, since color variations can get lost if a black-and-white copy of the paper is made.

For each font you have a choice of point sizes. In the early days of metal typesetting, there were 72 points to an inch. Thus, a 12-point font has maximum vertical letter size of 12/72 or 1/6 of an inch. The maximum vertical size is measured from the lowest point of letters that have *descenders* (g, j, p, q, and y). You should normally not use a point size less than 10 nor larger than 12 (except for titles). Of course the higher the point size, the less material fits on one page. Some people have suggested that we all use 12-point font size to ensure readability.

Page Attributes

Page attributes include:
- orientation
- margins
- columns
- page numbering.

Orientation refers to whether the pages are the standard vertical (portrait), or they are turned sideways (landscape). Landscape is used for wide tables, but be aware that the reader has to either turn the document sideways, or move to the side to read these pages.

Within the same instruction set, most word processors allow you to set the page size. The standard is 8.5 by 11 inches.

Margins are an important property of a document. The default is a one inch margin on all four sides. However, if your pages are being bound on the left, you may wish to increase the size of the left-hand margin (be careful if you are printing on two sides – you would then need a different margin for odd and even pages). If you are struggling to fit a document on a single page (such as a resume), you may try decreasing the margins slightly. Don't overdo this, however, since you don't want the page to appear cluttered.

b. 7 methods was used to identify potential pollutants in the vehicle's exhaust.

c. Team members meet weakly to review progress on there project and it's issues.

d. Me and the architect reviewed the requirements and plan's for the new convention center.

e. Its easier to review requirements and change specifications first rather then fix problems after the bridge is built.

f. Dr. Henry and his staff looks at each of the requirements carefully before approving and put on the list.

g. If you have less than 5 notebooks they can be layed on the desk.

h. Both companys believe in the motto: "The data doesnt lie!"

i. Its not good to release software before testing according to it's requirements.

j. The system is compromised of 3 major components: input processing and output.

2. Name five types of grammatical errors that are most noticeable to you. Are there any grammar rules that you find difficult to master? If so, can you think of a way that will help you to remember it?

3. Name five words that are easily misspelled. Are there any that you find difficult to remember? Can you think of ways to help you remember?

4. What are the advantages of using spell checking software? Are there any disadvantages? Are there any mistakes that are overlooked by spell checking software?

5. Find three websites that contain rules of grammar. Rank them from best to worst. Why did you rank them as you did?

6. When enumerating items in a list, both the form: "a, b, and c" and "a, b and c" are acceptable. Which do you prefer? Why? Is there any difference in clarity between the two?

7. What do you think is the easiest font to read? What point size do you like? Why?

ensure that the reader can follow the logic of your paper. Remember that your ultimate goal – whether persuasive, explanatory, or otherwise – depends on being clearly understood by the reader. Start your outline with an introduction that discusses all the particulars: who, what, where, when, and why. Follow the introduction with the topics you wish to discuss, organized in a logical way along with subdivisions as necessary. An example is the following outline for a project plan for a new corporate network:

Project Plan
1.0 Introduction
2.0 Project Requirements
3.0 Project Approach
3.1 Software
3.2 Hardware
3.3 Network
4.0 Implementation Plan
5.0 Management Plan
5.1 Roles and Responsibilities
5.2 Reporting
5.3 Issue Resolution
6.0 Schedule
7.0 Project Costs
8.0 Assumptions
9.0 Summary

As you can see, the outline uses a top-down approach that gives the reader the big picture and then works down to the details. Everyone involved in this project from the manager to the implementation team members can read the plan and understand what it means to him or her. If possible, include a summary that will reiterate the major issues, such as benefits and management approach.

If the document will be written by a team, joint development of the outline is a good way to get ideas and buy-in from the team members – not to mention possible writing assignments.

Phase 3: First Draft – Now get writing!

Your goal for the first draft is to fill in the sections of your outline with words. This doesn't have to be done in the order of the outline – you can start writing wherever you wish. Writing the introduction can be a good way to get started since you will be articulating the ideas that you had during your planning phase. In the introduction, mention the purpose of the document, describe the audience, talk about the uses of the document, etc. Alternatively, writing the introduction can be a good way to finish, especially if you are writing about something that you are learning about during the writing process. In order to write some sections you may have to develop graphs, charts, or timelines in order to start the writing. If you know that you will be inserting figures, leave a space in the document for them. Having a figure will make the writing easier because you will simply be describing or interpreting it for the reader.

Remember that while writing a document can be compared to a construction project, it is much easier to make changes in the document than in the physical construction project. Moving sentences and paragraphs is much easier than moving walls! When you are writing your first draft, just let the words and sentences flow. They may be fine, you may decide to change them later, you might want to move them, or you might decide to delete them. In the writing process, getting into the mode – or flow – of writing means that you will be creating the building blocks for the document as the words flow onto the page. You should not worry about being perfect or complete at this point, just get your thoughts onto paper – you can edit it later.

Writing Strategies

It can be difficult to write when the words don't come easily. If you are fighting against writer's block, [Drawing here?] try the following strategies:

- Jot it down: If you are stuck and just can't get started, try just writing sentences, phrases, or bullets into the sections of your outline. They will spur your creativity and help you come up with more to say.

[C] We proceeded to utilize the strain gauges to map the variations across the entire length of the test apparatus.
[S] We used strain gauges to measure strain across the entire length of the test apparatus.
[A] The phrase "proceeded to utilize" is too wordy. Just say "utilized," or reword the whole sentence.

[C] It was my assigned task to walk over to the light switch and effect an action to extinguish the room lights.
[S] I was assigned to turn off the lights.
[A] This person must have had an assignment to write a 500-word essay and ran out of ideas at 400 words. This sentence is full of fillers. Phrases such as "It was my assigned task" disguise the subject of the sentence. Who is doing what to whom?

[C] Our professor announced that she was giving us a homework assignment that involved researching fuel cells and using the results of this research to write a 500-word essay.
[S] Our homework was to write 500 words about fuel cells.
[A] Again, the original is far too wordy. Our professor gave us an assignment, so why say she announced that she was giving one? And all of this can be succinctly conveyed by "our homework."

[C] The group will decide the question of whether more office space is needed.
[S] The group will decide if more office space is needed.
[A] You don't decide a question. You answer the question.

[C] The flashing green light is indicative of the vehicle being ready to function.
[S] The flashing green light means the car can be driven.
[A] The phrase "is indicative of" is wordy. How about just "indicates" or even simplify the whole sentence.

[C] The circuit board is intended to be built before the end of the year, while the project is not required to be completed for several years.
[S] Our team intends to build the circuit board before the end of the year. The company has several years to complete the entire project.

[A] "Is intended to be built" disguises the subject and verb in this sentence. What action is being performed and who/what performs it?

[C] *It was the resistance of the wires that caused heating of the electronic system.*
[S] The wire resistance caused heating of the system.
[A] "It was the resistance of the wires" is a very wordy and complex way to state the subject of the sentence.

[C] *We purchased a computer for the purpose of increasing the efficiency of the department.*
[S] The computer increased the department's efficiency.
[A] The phrase, "for the purpose of" is too wordy and could be simplified to simply "to." So we could have said, "We purchased a computer to increase department efficiency." If the actual action of purchasing is not the key point, you can simplify further to what is shown above.

All you need to do is look around you to find additional examples of overly wordy statements. You can look at documents you have written, or you can view the work of others.

The important thing in simplification is to eliminate words and phrases that do not add anything to the meaning. We caution you that there are some exceptions. If you were writing a textbook, a certain amount of repetition is needed to reinforce the learning process. If, for example, you were describing Newton's second law, you might write something like:

Newton's second law states that

$$F=ma$$

The applied force is equal to the product of the mass of the object and the acceleration. For example, if mass is doubled, the force must double to achieve the same acceleration. Alternatively, if force is doubled and mass remains the same, acceleration will double.

We could have ended this discussion of Newton's second law after the equation and still maintained all of the essential information. But the added descriptive material in the paragraph above can help a reader absorb the concepts. Some readers also need breathing space so they have

[B] Soon after you begin your university studies, you should try to visit the Department Chairman.

[N] Soon after you begin your university studies, you should try to visit the Department Chair (or Chairperson).

[Note: If you are talking about a specific Department Chair who happens to be male, you can refer to him as Chairman. Also, if you get clear signals from an individual indicating they prefer the incorrect traditional language, you should honor their wishes. For example, some women board chairs refer to themselves as "Chairman of the Board."]

[B] A man and his dog go for a walk together.
[N] A man and his dog go for a walk together.
[Just testing you. Since we refer to a man, gender specific language is perfectly acceptable.]

Of course, if you are discussing a particular individual, then it is acceptable to use gender terms. You certainly would not say, "Dr. William Jones is director of our laboratory. He or she meets weekly with the engineers."

One way to achieve gender neutrality is to use "he or she" or "s/he." While this is an easy way out, it can get quite tedious and overuse shows a lack of creativity.

Word processing programs can sometimes help you modify gender-specific language. For example, WORD will check this as part of its grammar checker. Just go to grammar settings and click on "Gender-Specific Words." The program will not only flag these instances, but by right clicking the mouse, you will get suggestions to make your writing gender-neutral.

It's sometimes challenging and time-consuming to modify gender-specific language to make it gender-neutral, but you will find it is worth the effort and you will improve with practice. Diverse audiences of people reading your paper will appreciate the time you have spent being more inclusive.

7.3 Headings

You divided your paper into sections when you developed its outline. You can also further subdivide the contents by adding new sections or headings. Headings should be well-chosen to accurately reflect the contents of the following section and should be parallel in structure. Dividing your document into sections and subsections with headings can structure your writing and help the reader follow the flow. In addition, a reader might be interested in a particular aspect of your document (e.g., the experimental technique, the conclusions, the analysis) and will be able to find that topic more quickly.

If your paper is short (five pages or less), there may be no need to divide it into sections and subsections. But if it is relatively long (more than about five pages), such subdivisions are recommended. The subdivisions take a form identical to that of an outline, with headings and subheadings of varying levels. The first level of the outline consists of the main categories of your paper. Then, if necessary, you can include one or more levels of subcategory.

The first level (often given the appropriate name, "Level 1") is all you will generally need to use. The headings for each of the sections are normally centered and use upper and lowercase as opposed to all uppercase. When writing a formal paper, you should use the same font and point for the Level 1 heading size as you do for the text. You should generally avoid having just a single line of a paragraph at the bottom of a page (with the paragraph continuing on the next page) or just the final line of a paragraph at the top of a page. Word processing software refers to these situations as "widows" and "orphans" since you are leaving the line by itself without any companions. You can even set up the word processor to prevent widows and orphans. The same applies to headings. You should never have a heading at the bottom of a page with the sections starting on the next page. It is better to have an extra space at the bottom, and have the entire section start at the top of the next page.

If you find it necessary to go to another level of headings (i.e., subdivide your major sections), the next level (Level 2) would normally be flush left, uppercase and lowercase, and italicized. So the differences between Level 2 and Level 1 are location and italics.

There are conventions for more levels of headings, but two levels should be more than sufficient for the writing you will be doing as a student. You should become familiar with examples of writing in your university or company and try to follow the accepted conventions. You will find variation in the way headings are presented. For example, Level 1 is sometimes completely uppercase. The guidelines we give above should be considered as a suggested default setting, assuming no other conventions have been adopted.

Numbering of headings is rarely needed in a technical paper. It is common, however, in textbooks with many sections and subsections.

7.4 Revising the Draft – Editing Techniques

Every time you look at something you have written, you will probably find areas that could be improved. So when do you finally call the project finished, or in the words of movie producers, when do you "cut…print"? The answer depends on the due date for your report or project. It also depends on your particular writing style, your level of experience, and the environment in which you are performing your writing task. If, for example, the environment is one where you have been given an assignment with a firm deadline and that deadline has arrived, you can no longer make further revisions. Some people make very few changes after the first draft, and others make extensive improvements.

You might be interested to learn that this textbook has gone through over a dozen drafts before being printed and bound. Of course this may be too many steps for most writing projects, which have stricter deadlines.

After the completion of the final draft, the editing process consists of correcting errors in spelling, punctuation, and grammar. With today's word processing systems, editing and revision take place as the document is written. However, at the end of the writing process, you need to take a good hard look at the final document in its entirety and make sure it is absolutely flawless. Use spell check on the word processing system carefully; as mentioned before, the spell checker does not know which word you mean to use, so read the suggestions and think before making a substitution. For example, "duly" and "dully" are both spelled correctly, but the meanings of the words are very different.

If at all possible, ask someone else to read through your document when you are done editing. At this point in the writing process, you have written, re-written, and read the words so many times that it can be very easy to overlook an obvious error. If no second reader is available, consider reading your document aloud, which can help you improve cadence and tone as well. (Your dog makes an ideal audience.) At the very least, take a break and look at your work one last time with fresh eyes before you decide it is ready to distribute. A document that contains errors – even minor ones – reflects poorly on the writer and is more difficult to read and understand.

Read your document out loud

To edit the document, some reviewers can just use the computer to make changes to the draft document. If you wish, you could use the edit tracking feature of most word processors. In the "track mode" option of the "Tools" pull-down, all additions are marked in a different color (and sometimes underlined), while deletions are presented in "line through" style (so you can still read the deleted text). Then you get the option of either accepting or rejecting any of the edits to create a final copy.

If you have any trouble editing at the computer, you should consider printing a copy of the paper, perhaps with increased line spacing, and then marking it up with a red pen.

Following are ten editing suggestions:

1. Reread each sentence of the report to make sure it is clear and well-structured. You might want to even read these sentences out loud. As with any editing, you want to try to look at this from the perspective of someone unfamiliar with the document. You know what you meant when you drafted things, so it is often difficult to spot sentences that are not clear.

2. Make sure that your document is logical and consistent. Check to see if you have clearly identified your major points.

3. Check for overly long and complex sentences that should be split into more than one sentence. Readers appreciate shorter sentences and often have difficulty following long, convoluted discussions containing many separate ideas.

4. Use the Thesaurus to find the best word in cases where you are not sure. The word processor contains a Thesaurus, and all you need to do is position the cursor on a particular word and click the Thesaurus button. This presents you with synonyms and allows you to find the best word or to avoid repeating the same word over and over again.

5. Avoid adjectives such as "great" and "good." Don't say things like "the results were great" or "the results were amazing"; these terms are vague' remember, you are writing as an engineer!

6. Choose a paragraph format and stick with it. For example, you may choose to indent paragraphs, or if the document is single-spaced, you may leave two spaces at end of paragraph.

7. Make sure when you use words like "its" and "whose" that it is clear what they refer to. Even experienced writers have trouble with this rule. For example, you might write, "The design group presented its results to the company executive board. Its reputation depends on the quality of the report". The problem is that the second "Its" could either refer to the design group or to the executive board. You need to clarify this ambiguity.

8. Check the position and description of all figures. Make sure you refer to all figures in the text.

9. Make sure to include an abstract and references if these are required components of the paper.

10. Check for grammar errors. The most common ones are summarized below.

Apostrophes:

- ✓ Make sure that you used an apostrophe for a contraction (which puts two words together): "it's," "can't"
- ✓ Make sure that you did not use an apostrophe for a possessive pronoun: "hers," "his," "its"
- ✓ Make sure that you used an apostrophe for a possessive noun: "cat's," "Jack's," "rocket's"

Commas, colons, semicolons

- ✓ Check for commas in lists: "software, hardware, and documentation."
- ✓ Check for unnecessary commas – they can change the meaning of the text.
- ✓ Check for needed commas –to break up a very long sentence or set off a phrase.
- ✓ Check for correct use of colons and semicolons: "The network comprises several components: servers, routers, bridges, and hubs." "The network is very complex; it comprises servers, routers, bridges, and hubs."

Contractions:

- ✓ Use apostrophes for blended words: "can not" becomes "can't."
- ✓ Be wary of "false" contractions: "a lot" is not a word!

Pronoun problems:

- ✓ Make sure pronouns are correctly used, especially compound pronouns: "She and her team will be there." "The final document will be signed by Marty and me."
- ✓ Make sure that your usage is consistent; for example, when you write about a company, do not refer to it as "they" in subsequent sentences.

organization and, although you may have done most of the work, it is a team and corporate product. Focus on making your document the best it can be and be happy to get additional suggestions and input. In fact, actively soliciting input and suggestions is a great way to get others to have a stake in the success of the effort – always a good thing!

Editing is not always fun. We all reach a point when we wish a project were completed. However, a relatively small amount of added time spent on editing can result in a much better paper, which will be easier for your audience to read and will enhance your professional standing. Another benefit of thorough editing is that your writing skills will continuously improve in both efficiency and quality.

7.5 Editing Tools

Microsoft Word has powerful editing features that allow you to track these changes. You begin by accessing the tracking feature. Depending on your version of Word, you may have to pull down the "*Review*" tab. If necessary, go to HELP (F1) and look for "track changes". If you track changes, you will see that the computer will mark the text as you make any additions, deletions, or other textual changes. . The deletions are highlighted in color and with lines through the letters. The additions are underlined and in color.

Why would you want to do this? You may change your mind and want to go back to a prior version. Also, if multiple people are reviewing a document, they can enter their suggested changes so that you can easily identify them. Then you can go through the document and either accept or reject each change. If you are happy with all of the changes, you can accept them all with one mouse click.

Stand-alone Programs

Besides the features built into word processing programs, there are also stand-alone grammar checking programs. *AllWrite!* from McGraw-Hill, is one of the more widely used programs. It not only checks a document and gives hints for correcting errors, but it also contains guidelines for effective writing. The software contains tests and practice exercises. This program might be of particular interest to ESL students, although we can all use help in our writing.

Some software is user-friendly, and other software has a long learning curve. Unless you use a program regularly, you will probably not become comfortable with its features and remember how to use it effectively. You might want to start by seeing if the features built into your word processor are sufficient for your applications. If not, you can investigate the other software available. If you are a student, don't overlook the help you can get if your university has a writing center (most do). Many such centers provide software for student use.

7.6 *Plagiarism*

Now that you are moving along so well in your writing, it's time for a word of caution. Suppose that one of your professors gave an assignment to write a research paper on fuel cells. You did a wonderful job of researching the topic and compiled a file of Internet articles mixed with technical articles from the literature. You then wrote your report, carefully keeping in mind that your professor said the paper must be in your own words. Where necessary, you reworded material from the original source, changing a word here and there and adding some of your own phrases. You prepared an attractive cover page, wrapped up the whole package with an overpriced plastic instant binder, and proudly submitted it to the professor.

A week later, you get the paper back with a red "F" on the cover and a note from the professor that you are to appear before the department discipline committee to answer charges of plagiarism. What a shock! How could this have happened to you? What's this whole plagiarism thing about? Well, we don't mean to scare you, but it is, indeed, a very scary subject. Intellectual property is carefully protected, and we all need to protect the rights of the person who creates intellectual property.

What Is It?

The word "plagiarism" comes from the Latin word for "kidnapper." If you do an Internet search to find the definition of this word, you will see numerous university websites listed. We will quote the statement from California State University, Los Angeles[2] since it is representative of the entire set of statements.

> *Plagiarism is a direct violation of intellectual and academic honesty. While it exists in many forms, all plagiarisms refer to the same act: representing somebody else's words or ideas as one's own. The most extreme forms of plagiarism are a paper written by another person, a paper obtained from a commercial source, or a paper made up of passages copied word for word without acknowledgment. But paraphrasing authors' ideas or quoting even limited portions of their texts without proper citation is also an act of plagiarism. Even putting someone else's ideas into one's own words without acknowledgment may be plagiarism.*

Note that the simplified definition refers to misrepresenting somebody else's work as your own. The definition is really very clear if we refer to exact copying of material that someone else wrote without giving full credit and presenting the words without quotation marks.

But how do you deal with paraphrasing or putting ideas into your own words? Isn't that what much of research is about? When you research, you locate previously-written materials about your subject and then summarize them in your own words. But by doing so, are you plagiarizing and therefore liable for negative actions (e.g., fail a course, pay a fine, have your reputation damaged)?

---

[2] Senate Policy Document, California State University, Los Angeles, <u>Statement of Student Rights and Responsibilities</u>, Amended September, 2000.

As with many things, it's a matter of degree and a matter of opinion. In some ways, this is parallel to patents. Suppose you have a machine shop in your garage, and you like the way a particular patented can opener is designed. You build your own using that design as a guide. Have you violated the patent? Strictly speaking, many would say yes. Are you liable? If the company holding the patent had a large legal department that currently did not have enough work to do, and the company wanted to make you stand out as an example to others, you might find yourself in court. But if you don't make multiple copies and sell them, chances are nothing will ever come of this since the company has only lost the profit for one can opener and they would have to prove that you would buy one of their products if you had not built it yourself.

If you paraphrase somebody's writing and submit it to your professor for a grade, you would probably not have a problem as long as you used your own words and acknowledged the source via either footnotes or bibliography. But if you paraphrase a famous speech, give it yourself, and then win a Nobel Prize, you might find the original author to be in a fighting mood. Using the Gettysburg address as an example (as we did in Section 7.2 while discussing simplifying), suppose you made the following change:

Original: *Four score and seven years ago our fathers brought forth on this continent a new nation, conceived in Liberty and dedicated to the proposition that all men are created equal.*

Your version: Eighty-seven *years ago our ancestors advanced the concept that in North America everyone believes in liberty and has the same rights as everyone else.*

Hmmm. You did a pretty good job. But the similar structure and ideas would probably make this plagiarism unless you included a reference such as, "Adapted from Lincoln's Gettysburg Address" Or added an introductory reference such as, "As President Abraham Lincoln noted in his famour "Gettysburg Address,"…

So as you do your research and writing, be sure to reference any idea that is not your own. In Section 8.2, we discuss referencing, which you should do even if you use your own words to convey an idea.

We warned you that this could be scary. If you are basing your paper on the work done and reported by others, avoiding plagiarism is challenging. This is especially true since paraphrasing the original document is not sufficient. You really have to put things completely into your own words and your own sentence structure. This requires that you fully understand the concepts in the original paper. Even if you do this, you still need to acknowledge your source!

Public Domain

Materials in the "public domain" are considered public property and available to anyone to use. Such materials are not covered by copyrights and you can include them in a paper without obtaining permission from the author or publisher. Public domain publications include those created by agencies of the United State Government, and other documents that are not eligible for copyright (e.g., most mathematical equations). Public domain also includes those documents for which the copyright has expired. Although public domain materials are available for use without securing permission, it is still good practice to reference such items if you use them in your documents.

Who Cares?

If you steal a candy bar and don't get caught, should you feel happy or sad? If you steal someone's intellectual property but nobody else notices, does anyone care? Of course, you should care. You have to live with yourself, and if you are claiming credit for work that others have done, we hope this will make you feel uncomfortable. Personal ethics cannot be taught in a course or through a textbook.

But if your principles of personal ethics are not motivating enough, just think about getting caught. If you are a student, the first level of policing rests with your professor. Experienced professors can easily tell when submitted writing assignments do not match the writing skills and styles of a particular student. Professors also have software and online tools at their disposal. An increasing number of sophisticated services exist for comparing submitted papers with the online literature to find patterns of matching. Universities are taking plagiarism more and more seriously, so this type of monitoring is sure to increase with time. If you want to really feel paranoid, take a look at http://turnitin.com/static/plagiarism.html. Apparently, a growing number of universities are subscribing to this and similar services. Corporations also have a major concern, since they are on the receiving end of legal action if ideas are copied.

Penalties at the university level depend on the severity of the plagiarism and on the approaches favored by the professor. Sanctions can range from a failing grade on the assignment to a failing grade in the course. At most universities, the professor also has the option (sometimes even the obligation) to bring charges against the student, escalating it to the university review level. At the university level, panels made up of faculty and students usually review the evidence and recommend corrective action or sanctions.

As your papers get wider and wider circulation (e.g., they are posted on a website), a much larger audience gets to see them. If someone reading your work has the feeling, "I've seen this somewhere before," it could have some very unpleasant repercussions. Indeed, even the original author from whom you plagiarized might end up reading your paper. That author would be the most likely person to recognize ideas that you have put into your own words without referencing.

The various professional organizations have strong positions on plagiarism associated with their many publications. For example, the IEEE defines plagiarism in its documents and encourages editors and readers to report suspected incidents. Following investigation, if plagiarism is confirmed, a number of levels of sanction are possible. A notice can be posted (in *Xplore*) for all to see. The plagiarist can be prohibited from further publishing in any IEEE journal or periodical, and the matter can be referred to the "Ethics and Member Conduct Committee" for further action.

Technical professionals must also conform to corporate policies. As mentioned above, companies take incidents of plagiarism very seriously because they can have far-reaching consequences. When you "borrow" others' ideas and/or words, be sure to give full credit to the originator. It's good practice and will keep you out of trouble.

We hope that this information frightens you a bit. Plagiarism detection techniques are bound to improve and sanctions to increase in the future, so you should make every effort to acquire good citation habits early in your professional career.

Plagiarism — How to Avoid It.

Gene Fowler, a noted journalist, authored many witticisms both spoken and written. Two regarding the art of writing are especially appropriate:

☐"Writing is easy. All you do is stare at a blank sheet of paper until drops of blood form on your forehead."

☐"The best way to become a successful writer is to read good writing, remember it, and then forget where you remember it from."

The first of these witticisms suggests that writing is a painful process. This directly contradicts the primarily principle of this text. Though writing does not necessarily come easily, it should not cause a great deal of pain.

The second witticism is almost scandalous. It would seem to encourage plagiarism. Of course, Gene Fowler was probably talking about general techniques of writing rather than specifics, so we will give him the benefit of the doubt. He certainly could not have been encouraging people to mimic the writing of others without giving full credit.

We have already explained that if you keep the same sentence structure as the original but merely reword phrases and substitute words, you are guilty of plagiarism. Even if you reorganize sentence structure, you are still on shaky ground even though you are making progress toward creating a paper you can call your own. You still need to cite sources of ideas, no matter how much sentence re-structuring you do!

If you do perform these various manipulations (i.e., rearrange sentences, substitute expressions and phrases) and simultaneously give credit to the source, you are probably protecting yourself against charges of plagiarism. But we hope you will want to go further to develop your own writing style and create a document that is truly your own.

There is a big difference between writing a paper that summarizes what others have done (and already reported on) and writing about something new that you have developed. In the latter case, plagiarism is usually not an issue since you are the first creator of the work. Likewise, if you take someone else's work and significantly enhance it and add your own material, you will probably have sufficient new content to avoid plagiarism. So we are really talking about the situation where a substantial part of the technical detail has already been written by someone else and you have the assignment of summarizing it as part of your document.

Authors develop their own techniques to avoid inadvertent copying. Most of these techniques require that you <u>not</u> have the original in front of you while preparing your document. So the practice of many people of cutting and pasting from the web and then trying to modify is ripe for encouraging plagiarism. It's made us all lazy. We suggest that you take notes while reading the original documents. As you take these notes, clearly indicate those parts that are direct quotes. You might take the original documents and use a highlighter pen to mark significant statements. Then summarize these highlighted statements in your notes. If you have enough confidence, take your notes from memory: put the original document away or close the web page before writing them. This will help ensure that you absorb the material before you start your own writing which will then be in your own words.

Once you have created a set of notes, look through them and decide where you want to add material or combine parts. Mark up the notes thoroughly.

Then use the final version of the notes when you draft your paper. By this time, you should no longer have a copy of the original document in front of you. If you have been careful to take notes and not simply copy full sentences and phrases, you will most likely end up with a document you can call your own.

Plagiarism is a great topic for conversation among your colleagues (a nice party topic?). Ask other students or professionals what they do to avoid plagiarism and how they feel about it. As is usually true, you can learn a lot from others.

7.7 Exercises

1. You have been assigned to write a proposal to build a new factory. Create a schedule for developing the proposal that incorporates input from the architect, client, and your management.

2. Following is a diagram for a standard computer system.

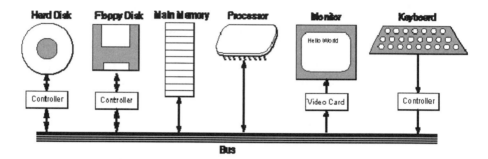

Main Components of a Computer System

Write a description of the figure.

3. Now imagine that the system is a new one with state-of-the-art components (expand). Write a description of the advantages of the new system.

4. This is a network diagram for a Local Area Network (LAN)

A Local Area Network (LAN) with Printer

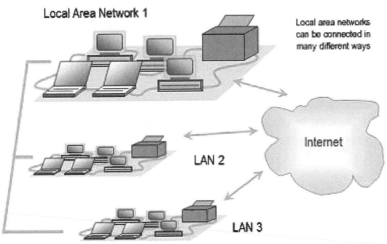

You are working on a proposal to install the system at three sites for a banking company. Develop a technical approach describing how you would install the system including a project plan, roles & responsibilities for your company and the client, etc.

What assumptions did you make? What questions do you have?

5. Rewrite the following sentences for gender neutrality:

    a. If a manager wants to attend the annual meeting, he should make sure that he has been approved.

    b. If the mathematics program is too difficult for an engineering student, he can obtain personalized tutoring from this office.

    c. An engineer should be aware of his professional responsibilities and make sure that he always completes his assignments on time.

    d. Since the receptionist wasn't at her desk, I took a seat to wait for her to return.

e. According to security rules, an engineer must lock up his
    work before leaving.

6. Which of the following sentences do you prefer and why?

7. An analyst needs to consider all potential alternatives before choosing
    the best approach.

8. Analysts need to consider all potential alternatives before choosing
    their favorite approach.

9. Edit the following introductory paragraph:

NASA Launch Decisions

Final decisions about NASA launches are made by the MMT. They
are supported by huge amounts of data to ensure all issues are
anticipated before a question arises that might jeopardize the launch.
The Mission Management Team review the data which is analyzed
carefully along with procedures which are followed through the
countdown period. If a problem is discovered that could jeopardize
the launch, the team has a meeting to review the severity of the
problem and its potential results. The decision to halt a launch is not
made lightly and many people are involved in the discussion. For
example, if there is a faulty fuel gauge, engineers need to be
consulted and try to make repairs on the launch pad if possible. The
MMT look at the factors necessary for a safe launch including the
launch vehicle, many countdown procedure, weather conditions and
planetary positions.

10. Without looking back at the chapter, list five tasks you think you
    would have to perform in revising the first draft of a paper. Then
    compare your list with the items presented in this chapter and discuss
    any significant differences.

11. Locate three university websites that address plagiarism. Compare the
    definitions and briefly discuss any differences.

12. Locate three university websites that discuss ways to prevent plagiarism. Summarize what is written in these three sites (without plagiarizing the material).

13. Search the web to find an example of plagiarism and write a short paper describing the incident and the outcome. This

14. Locate an article in a journal in your field (e.g., ASCE, ASME, IEEE, ACM). Rewrite the first paragraph of this article in your own words.

15. Interview three of your professors or, if you are not a student, interview your supervisor. Find out their opinions about plagiarism and how they would deal with such incidences.

informative abstract talks about the details of the *subject* of the paper It includes results presented in the paper and summarizes each section. For brevity, it does not present introductory material and omits extensive detail.

In industry, an informative abstract is called an executive summary and can be longer than the abstract of a scientific paper. An executive summary is used to summarize a long document such as a proposal succinctly in a few pages. Higher-level managers and executives may not have time for or interest in all the details in the document, but need to know important facts for management and decision-making.

A common guideline for the length of an informative abstract is 10% or less of the length of the paper. If you are the type who uses a highlight pen to select key points within a paper, you can think of this abstract as a summary of what has been highlighted (using full sentences instead of phrases).

An informative abstract of this section of the text might look like the following:

> An abstract summarizes the key points of a document. It is meant to give the reader guidance about what is covered in the full paper. Abstracts are usually attached to the paper, although they can be separated and used in such applications as abstract indexing. There are two types of abstract: descriptive and informative. The descriptive abstract gives an overview of the paper without giving any details about the topic. Alternatively, the informative abstract summarizes the key results presented in the paper. If the abstract is informative, the writer need simply summarize each major point or paragraph. If the abstract is descriptive, the writer must answer the question, "What is your paper about?"

How to Write an Abstract

You usually cannot start writing an abstract until the paper is completely finished. The only exception would be if you write a rough draft of a descriptive abstract that is then used as a planning document for writing the paper. That is, you can use the descriptive abstract as the core of a suggested outline or planning document for writing the paper.

A descriptive abstract is normally quite easy to write. The main thing to do is familiarize yourself with the style of writing used in these abstracts. Because you have written the paper, you should be quite familiar with the paper's contents. Try to imagine describing your paper to a classmate or colleague in a short memo. Suppose you are asked, "Describe your paper." How would you respond? Write down your thoughts and then refer to your paper to see if you have included everything. Some elements of a descriptive abstract are:

- The thesis of the paper
- The intended audience
- The purpose
- Methods and results.

Jot down your thoughts and use them to write a short descriptive abstract (usually fewer than 100 words in length).

Now suppose you are writing an informative abstract instead of a descriptive one. Because the informative abstract summarizes the entire paper, you should go through the complete paper paragraph by paragraph. If you have used a highlighter pen, writing the abstract will be easier, since you only need to focus on the phrases you have highlighted. Then summarize each paragraph or section, keeping the 10% length guideline in mind. If you use the exact sentences from your report, you will include too much information, so be sure to summarize. As with the descriptive abstract, we strongly recommend that you look at many samples from the literature to acquaint yourself with accepted styles of writing.

Elements of an informative abstract are:
- The thesis or purpose of the paper
- The intended audience
- Scope
- Methods
- Results
- Recommendations or next steps

volume number, and (optionally) the page number(s). If some of these cannot be found, you need not lose sleep. The components of a reference can have different formats. For example, we have used italics for the title of the article, but in some cases, the title is enclosed in quotes. In such cases, the publication title appears in italics or is underlined to distinguish it from the article title. Give as much information as you can find. Here are some examples:

Noll, P., *Wideband Speech and Audio Coding*, IEEE Comm. Mag., Vol 31, No 11, Nov 1993, pp 34-44.
Birch, Stuart, *Fuel cells edge closer,* Automotive Engineering November 1997: 102

Electronic Sources

In referring to electronic sources such as data files and websites, your goal is to provide enough information so that the source can be located. It is often impossible to find all of the information you might like on a website such as the author's name and date of publication.

You should try to include the author/editor, the year, the title, and type of medium in your reference along with the company or creator of the material and the web address. In some cases it is helpful to also include the month, day, and year when the information was retrieved. This is particularly important if a website changes frequently.

Here are two examples of web citations:

Australian Bureau of Statistics (2000). *1996 Census of Population and Housing: Northern Queensland (Statistical Division).* [Data file]. Available from the Australian Bureau of Statistics site: http://www.abs.gov.au

Jones, G. P. (2001). *Holidays on the Net.* Retrieved October 21, 2001, from http://www.holidays.net/.

If you did web research on a subject such as high-definition radio, you might quote some material from the "How Stuff Works" website. The reference for this would look like:

Grabionowski, Ed. "How HD Radio Works". *How Stuff Works website*: http://electronics.howstuffworks.com/hd-radio2.htm

There are two observations we want to make about this example. First, although it may appear that the author edited this (see the "Ed" after his name), that is not the case. His first name is Edward, and he apparently goes by Ed. But also assuming that the webmaster knew all the rules and that Mr. Grabionowski was actually an editor, the "ed" would be lower case.

We also could not find any date or location of the article, so that information is missing. We could probably find out where *How Stuff Works* is based, but perhaps it is someone's garage, so we would be better off not knowing!

Be sure to test every URL you reference in your paper to make sure it is accurately copied. Of course, that still does not assure that the URL will still be active by the time someone reads your paper. How many times have you tried to go to a web pages listed in a textbook or article only to find "Page Not Found"?

You need to make every effort to give references for material generated by others. This is governed both by professional ethics and by law. So even if you had conversations with an engineer in industry, you should document as much as possible. If, for example, you had contacted Raymond Landis to discuss student study habits, you should include something like:

R. B. Landis (personal communication, July 15, 2008).

Similar rules apply to referencing emails. You should include the sender (sender's e-mail address). (year, month day) and the subject of message. You have a lot of freedom regarding organization of the material in this type of reference. Here is one example:

Suzuki, Keiko (keiko@hotmail.com). (2001, Nov. 2). Traditional foods during the holidays. E-mail to Tom Jones (jones@gol.com).

Placement of References

Let's take a look at the placement of references within a document. We will select a paragraph from this text and give three examples of ways that references can be added.

publications on a particular topic. In these cases, the creator of the bibliography does not necessarily need to have accessed each publication and used it in preparing the paper.

In contrast to these comprehensive bibliographies, most of the ones you write should be limited to resources you have seen and used in preparation of your paper. Try to avoid esoteric entries (e.g., "Proceedings of the 1968 International Conference on Widgets for Composite Material Pistons held in Estonia") unless you have actually found these in the library and can honestly say they are relevant to the topic of your paper. The listing is useless if the average reader of your paper is not able to obtain a copy of the referenced work. Try not to think of this as a place where you brag about how extensively you conducted research.

Bibliographies contain a variety of resources including books, magazine and newspaper articles, and Internet resources. This has become the general practice, even though the original definition of bibliography referred only to books.

Most of us dislike adhering to rigid guidelines. But when it comes to bibliographies, it is advantageous to have certain standards for what is presented and in what order the details are given. If you have a strong aversion to reading a whole list of instructions, you can instead try looking through your library to see examples of bibliographies. The first place to look would be the textbooks you are using for class. You can assume the authors and publishers have used consistent and accepted formats and styles. Of course, if you are writing a paper in response to a class assignment, make sure you know the rules of your university or your particular professor. The university rules can typically be found at the university writing center or on the center's website. You can also look at examples of previously submitted papers.

The general rules for listing resources are the same as those for references. That topic is covered in Section 8.2.

8.4 Figures

Graphics help illustrate viewpoints

It is said that a picture is worth a thousand words. In fact, in technical communication, diagrams can be worth far more than that. They can help the reader visualize concepts and put things into perspective. They can also help turn a boring paper into a more interesting one. Nobody likes reading paragraph after paragraph of text without anything to break it up. Visual aids can include line drawings, graphics, bar charts, pie charts, and even photographs.

There are many stand-alone software programs available for making drawings. Among the most popular are Microsoft Visio (and Visio Technical) and SmartDraw. You can also use Adobe Photoshop to modify photographs and drawings, or you can use Excel to create graphs, bar charts, and pie charts. If your primary interest is electrical circuits, you can use Electronics Workbench or OrCAD. Many SPICE-based programs (e.g., MicroCap and Tina) can also create professional-looking circuit diagrams. Most word processing programs also have built in drawing capability. Microsoft Word can be used to draw objects and charts, as well as import clip art both from the program files and from the web.

The problem with any drawing program is the learning curve. Unfortunately, no drawing program is completely user-friendly. Unless you plan to use the software on a regular basis, you may not have the incentive to learn all of the sophisticated features. Another problem is that much of this software is intended for design use and therefore is far more powerful than necessary if all you want to do is draw diagrams. You should purchase a piece of software only if you plan to use many of its features in the future.

Once you have drawn diagrams using any of these programs, it is a simple matter to import them (or copy and paste) into a word processor document. If you have many diagrams in your paper, you may want to consider numbering them for easy reference. If you do number figures, you should number every figure even if you don't reference a particular one in the paper. This makes it easier for the reader to find a particular figure. You should learn how to add captions to your drawings. You should always refer to diagrams in the text of your paper. If you include a diagram without any reference, the reader may read the paper without examining the diagram. At the least, you should have one sentence saying something like, "The following diagram [or Figure xx if you are numbering] shows test results for our design."

Microsoft Word has drawing features such as creating simple lines, arrows, rectangles, and ovals. Depending on the version of the software you are using, you access the features in various ways. Use the "help" screens (F1) to guide you. You can draw shapes, diagrams, flowcharts, curves, and lines; or you can use WordArt.

Try inserting artwork from a web page (be careful of copyrights – give full credit for anything you use). Also try using color in your diagrams, although you need to be sensitive to the fact that many papers you write will be duplicated in black and white. It's tempting to use multi-colored bars on a bar chart, but then those who get black and white copies have no way to distinguish which is which.

A word processor can be used to draw various types of charts, including bar charts. However, a word processor does not do any data analysis and only draws what you tell it to.

Excel creates graphs and charts using data that you input. Let's look at the following example presenting population information in four different formats:

Text format

The population of AnyTown was 20,050 in 1960; 23,065 in 1970; 27,000 in 1980; 31056 in 1990; 36,432 in 2000; and 40,123 in 2010. If you do this in tabular form, it looks like:

Table format

Population of AnyTown, USA

1960	20,050
1970	23,065
1980	27,000
1990	31,056
2000	36,432
2010	40,123

Line graph format

[see next page]

Creating Tables in Microsoft Word

If you want to add a table directly in Microsoft Word (procedures are similar for other word processors), simply position the cursor where you want to draw the table. Then click on *Insert*, and then select *Tables*. Depending on the version of Word you are using, the choices will be presented in different ways. You can create a standard table by either specifying the number of rows and columns or by dragging the mouse across a template to set the size of the table. Alternatively, you can draw a table from scratch. When you make this drawing selection, the cursor pointer changes to a pencil, and you click and drag to draw the outside boundaries (a rectangle). Then you click and drag to add each horizontal and vertical line. You can erase lines by clicking *Eraser and then clicking* on the line you want to erase.

Here is an example of a standard table with four rows and five columns:

Note that the widths and heights of the cells are uniform throughout the table. Once the basic table is drawn, you can move the vertical lines around to create different width columns.

We now repeat this example using the *Draw* feature to make cells of differing sizes.

Once you have the table drawn, you can enter data directly. You can also insert graphics in any cell. You can merge cells and also change individual borders. Here is another an example of the first table with data and titles entered.

Date	Experiment #	Test Results for two tests		Comments?
1/3/08	1	21.5	21.7	No
1/4/08	2	22.3	22.1	No
1/5/08	3	24.1	22.1	Yes

Table 1 - Test Results

Notice that we have changed the point size of the headings and used bold type, merged two cells of the headings, added a caption, and put a different border around the top row entries of the table. Most of this can be done by highlighting the cells and right-clicking the mouse to see the various selections. We hope you agree that these easy changes have made the table more interesting and readable.

Creating Tables Using Microsoft Excel

If you are not already familiar with creating tables in Microsoft Excel, we recommend that you take the time to experiment with the software and learn the basics. Aside from the very powerful features that allow you to manipulate data, Excel enables you to format the table in different ways. You can choose from a number of templates for table design, or you can modify features individually. These changeable features include fonts, merged cells, borders, and cell sizes.

This text is not intended to substitute for a software manual, so we have not included specific individual instructions. While in Excel, you can always get online help by pressing F1.

Once you have the table looking the way you want, you can simply import it into your word processor. You can insert the Excel file, but it is often easier to just select the table and then copy and paste into the document. If at all possible, avoid splitting a table onto two pages. It is best if the reader can see the entire table at once.

8.6 Equations

As a technical professional, your writing may contain equations. Creating well-designed equations requires time and practice. You should try to learn to write equations with the word processor program. However, should you find this too difficult (it can be cumbersome), you may wish to investigate the use of software such MathType[8]). Some word processor software uses equation editors that differ significantly from that of Microsoft Word. In its basic form, the Word equation editor requires you to click many icons to create complex equations. Built-up fractions are entered by typing "x over y" which creates

$$\frac{x}{y}$$

Most current word processor programs have adopted the Microsoft Word technique for entering equations, so we recommend that you learn that. As with any software, the best way is to start with very simple equations and then add features as needed. If you try to learn all

---

[8] MathType is a powerful interactive tool for Windows and Macintosh that lets you create mathematical notations for word processing, web pages, desktop publishing, or presentations.

the features at once, you will forget the earlier ones before you even apply them.

To add an equation to your document, position the cursor and then click *Insert* followed by *Object* and select *Microsoft Equation X,Y.* Some newer versions of Word allow you to go directly to the "Equation Editor" from the *Insert* menu. If you use this editor a lot, you can add a selection to your tool bar to permit going directly to it. When you enter an equation using this editor, Word automatically converts the equation into one that is professionally formatted.

Word has a variety of built-in equations. If you use a particular equation on a regular basis, you can add it to the list. You also can start with something on the list and then modify it.

You can always get assistance from the "help" screens (F1). Alternatively, you can use the web to find many excellent tutorials created at universities around the world[9].

Equations are usually centered between the left and right margins. In some instances, a very simple equation can be inserted in the middle of a sentence without giving it a separate line, but this is rare and could make it more difficult for the reader to follow. Equations can be considered as part of a sentence so you might write something like the following:

The relationship between force, mass, and acceleration is given by

$$f = ma.$$

Note the period at the end of the equation. Punctuation (periods or commas) is used just as if you were writing a sentence. This rule is not always followed strictly, so check some publications from your university or employer to see what conventions they use.

---

[9] The University of Waterloo has a comprehensive web site giving not only excellent tutorials but also video clips demonstrating the equation editor features. Its address is: http://ist.uwaterloo.ca/ec/equations/equation.html.

Numbering Equations

If the document you have written is relatively short and has just a few equations, there is usually no need to number them. You can refer to them using descriptions. But if the number of pages or equations is large (say more than ten), numbering is desirable. Even if you don't need to refer to the equations by number, a reader of your paper may wish to do so. Some writers only number those equations that are considered significant and that may be referenced by others. However, it is often advantageous to number every equation. This helps the reader locate a particular equation more easily. If there is just one equation number every several pages, the reader may have to struggle to find a particular number.

Numbering can be done sequentially using (1), (2), (3) and so on. Alternatively, if your paper is divided into sections, you can use a layered equation numbering system, such as (1.1), (1.2), (2.1), (2.2), etc.

The best situation would be to be able to configure the word processor to automatically insert numbers. Then if you later add an equation, you don't need to manually change all of the later numbers. Traditionally, Microsoft Word has not provided this capability. If you use a search engine to find articles on numbering equations, you will find tutorials from a number of universities. Be patient and you will eventually get the hang of it.

The process of numbering equations is sufficiently cumbersome that some people "cheat" and make the equation number part of the actual equation. They tab after the formula portion and then "right-justify" the entire line. This latter technique is awkward, and it is difficult to line things up properly. So it is best to learn a technique or wait until Microsoft solves this problem (perhaps even by the time you are reading this).

8.7 Exercises

1. Select a book or article in your field and write an abstract for it. Is your abstract descriptive or informative? How difficult was it to write the abstract?

2. What are the problems with using web-based material for references? What should you do if you need to cite these materials? What should you do if you find out that the website is run by a student?

3. What would your reference look like if you used this book as a resource?

4. Give some examples showing the differences between books used as references and those in a bibliography.

5. Develop a table of data and the corresponding bar and pie charts.

6. If you are systems professional, you may need to develop flow charts. Research available software tools for flow charting and write a recommendation for your organization's use. What are the criteria you used for your recommendation?

INDEX